# Locomotives of the Somerset & Dorset Joint Railway

**Loco crew** and a selection of station and and other staff happily pose in front of a highly burnished 1861 built George England 2-4-0. The date is unclear, but this loco was originally numbered 2, but became 25 in June 1876 then No. 25A in August 1881. So the picture was taken in this five year period. No. 2/25/25A was withdrawn from service in May 1885. (D Neal)

# Locomotives of the Somerset & Dorset Joint Railway

## A Definitive Survey, 1854–1966

**TIM HILLIER-GRAVES**

First published in Great Britain in 2021 by
Pen and Sword Transport
An imprint of Pen & Sword Books Ltd.
Yorkshire - Philadelphia

Copyright © Tim Hillier-Graves, 2021

ISBN 978 1 52674 835 5

The right of Tim Hillier-Graves to be identified as author of this work has been asserted by him in accordance with the Copyright, Designs and Patents Act 1988.

A CIP catalogue record for this book is available from the British Library.

All rights reserved. No part of this book may be reproduced or transmitted in any form or by any means, electronic or mechanical including photocopying, recording or by any information storage and retrieval system, without permission from the Publisher in writing.

Typeset in Palatino by SJmagic DESIGN SERVICES, India.
Printed and bound in India by Replika Press Pvt. Ltd.

Pen & Sword Books Ltd incorporates the Imprints of Pen & Sword Books Archaeology, Atlas, Aviation, Battleground, Discovery, Family History, History, Maritime, Military, Naval, Politics, Railways, Select, Transport, True Crime, Fiction, Frontline Books, Leo Cooper, Praetorian Press, Seaforth Publishing, Wharncliffe and White Owl.

For a complete list of Pen & Sword titles please contact

PEN & SWORD BOOKS LIMITED
47 Church Street, Barnsley, South Yorkshire, S70 2AS, England
E-mail: enquiries@pen-and-sword.co.uk
Website: www.pen-and-sword.co.uk

or

PEN AND SWORD BOOKS
1950 Lawrence Rd, Havertown, PA 19083, USA
E-mail: Uspen-and-sword@casematepublishers.com
Website: www.penandswordbooks.com

**Contents Page Image:**
**The heyday** of the Somerset and Dorset Joint Railway is captured in this photograph of Highbridge Works at the end of the nineteenth century with a cross-section of the engines that then populated the line. (Author)

# CONTENTS

| | | |
|---|---|---|
| | Acknowledgements | 6 |
| | Introduction | 8 |
| Chapter 1 | A Line Created and Remembered | 12 |
| Chapter 2 | Locomotives in the Early Years (1854–75) | 48 |
| Chapter 3 | The Midland Years Part 1 (1876–1903) | 74 |
| Chapter 4 | The Midland Years Part 2 (1904–22) | 120 |
| Chapter 5 | The LMS Years (1923–47) | 155 |
| Chapter 6 | A Long Goodbye (1948–66) | 200 |
| | Reference Sources | 239 |
| | Index | 240 |

# ACKNOWLEDGEMENTS

During my life, I have been privileged to meet many people who worked on the railways and were happy to paint vivid pictures of the days when steam dominated their day to day lives – engineers, designers, footplate crew, managers and many more.

Their numbers grow thinner each year, but luckily many have been moved to reminisce or write about their experiences. This has created a bank of information for future generations to enjoy and so increase their understanding of a time of which many can have no personal knowledge. To this we can add many other individuals who have made a huge commitment, personal and financial, to preserve and record so many aspects of railway history. Amongst these most dedicated people I number my late uncle, Ronald Hillier, whose legacy

**I think** this photograph taken at Burnham station in 1895, according to notes written on the back of this old and badly faded print, sums up the character of the S&D very effectively. A rural line in a firmly established Victorian country where the pace seems to be relaxed to say the least without the bustle or pace of city life. This will soon change with the arrival of a new century, two world wars, the growth of consumerism, the rapidly growing domination of the internal combustion engine and the hard economics of the modern age. All this will sweep away the world captured in this picture and with it the S&D. The locomotive is No. 28, an 0-6-0 tender engine built by the Vulcan Foundry in 1881. It survived in service until 1928. (Author)

has helped me to write about many railway related subjects, including the Somerset and Dorset (S&D) over which he travelled on many occasions.

No list of contributors would be complete without my old friend David Neal. In the Fifties and Sixties, when so much was being destroyed, David collected many unique items relating to the S&D in all its guises. He sent me copies of these and gave me permission to use them as I saw fit.

National and local institutions also played a huge part in this process. The National Railway Museum and National Archives sit at the centre of this work, but they are ably supported by many other bodies, including many preserved railways which have created their own libraries and collections. We are very lucky in Britain to have so many sources of information to inform our history.

The principal sources of material I have used are listed in the reference section of this book. Occasionally, I have quoted directly from these items, mostly because I couldn`t improve on the descriptions or the technical assessments made by specialists in these fields. I thank the authors or holders of this material for permission to use it to enliven this book and give it greater authenticity.

To all the people who helped me write this book, I give my thanks and hope I have done justice to all that they have contributed. Ultimately, though, all an historian can do is to sift and consider all material and reach a judgement that he or she thinks honestly represents history. There will, undoubtedly, be alternative views or conclusions and even some omissions that someone thinks may be crucial to a story, but that is as it should be. I don`t think there`s ever a final word and new material may be found to allow fresh interpretations to be made.

In producing photographs for this book, preservation work has been necessary. In some cases, their sepia finish, foxing and dilapidated condition could not be entirely overcome. However, because they are often rare pictures or have some historic significance they have been included despite their condition. I hope this doesn't spoil your enjoyment of the book.

Copyright is a complex issue and often difficult to establish, especially when a photo or item exists in a number of public and private collections, all of which could, in theory, claim ownership. Most material in this book was provided by the people listed and copyright has been assigned to them where appropriate. Rigorous checks have been made to ensure that each picture or item has been correctly attributed, but no process is flawless, particularly when these things are now more than 70 years old and the originators probably now long dead. If an error has been made it is unintentional. If a reader wishes to affirm copyright, please contact the publisher and an acknowledgement will be made in any future edition of the book, should a claim be proven. I apologise in advance if a mistake has been made.

# INTRODUCTION

When asked to write about the Somerset and Dorset Railway by Pen and Sword's commissioning editor, I leapt at the chance. Like many railway enthusiasts, I have long been intrigued by this unique line. In my case, awareness of its existence came in 1956 when taken to see the Ealing Comedy *Titfield Thunderbolt*. The film opens with a short sequence showing an express pulled by a Bulleid Pacific on the S&D passing over the star of the film – a GWR 0-4-2 tank engine on the imaginary 'Titfield' branch line. It is only a second or two of the film, but is memorable, nonetheless.

A few years later my family moved from London to Bath. Each day, whilst going to school, I passed Green Park Station, the company's Bath terminus. Here I saw for myself the Somerset and Dorset in action. Sadly, by this time the line was soon to become another Beeching cut and the end was in sight. However, there was just enough going on to catch my attention, although the usual smell of steam and smoke from the engine sheds was overlaid with less pleasant odours emanating from Bath's gasworks nearby, now also a distant memory.

I often had the opportunity to wander around the station and sheds and see what was going

**A postcard** produced in great numbers that seems to capture the essence and individuality of the Somerset and Dorset Joint Railway – an elegant large-boilered Deeley 4-4-0 painted in Prussian Blue. (Author)

on. In those days, security was minimal, if not non-existent, and you were more likely to receive a pleasant greeting from BR staff than get a ticking off for trespassing. So, until closure, I was able to witness many things, including its death throes. Sadly, I never had the funds to travel southwards across the Mendips to Bournemouth and experience other aspects of this fast disappearing world. But with my Raleigh drop handled bike and a borrowed Leica llf camera, I did journey down the line to watch and photograph many engines as they passed by. Then it was gone, with only Green Park Station remaining as a reminder of past glories. This might also have been demolished but for the efforts of conservationists and the construction of a Sainsbury's store which was integrated into the old station.

Fast forward again to the 1970s and the appearance of Ivo Peters' S&D themed books displaying a small part of his immense photographic collection. Thanks to his dedication, life on this line over the last two decades of its existence was captured in great detail, but, more importantly, by someone in tune with all its foibles and charms. It is an incomparable record which is treasured to this day. But others also helped capture the essence of this world and here three other books added greatly to my appreciation of all it stood for. These are *Footplate Over the Mendips* and *Mendips Engineman* by Peter Smith and Robin Athill's *The Somerset and Dorset Railway*.

I think it impossible to read these books and not be captivated by the company's history, the character of this railway and not be moved by its loss. It was worthy of much more and if saved might have found a role for itself in our modern world, especially at a time when internal combustion faces its own demise due

**The eleven** Midland Railway built 7F 2-8-0s served the S&D very effectively throughout their long lives and represented all that this singular line represented. Here engine No. 53804 pulls a load of trucks into Evercreech Station in the early 1960s, with another symbol of the time, an Austin A30, parked nearby.

**On 13** October 1964, with the grandeur of Bath's Georgian buildings as a backdrop, the S&D's sheds still ply their trade with Stanier 8F, No. 48309, soon to be on duty. These engines were thought by some to be ideal for use on the demanding hills to the south of the city and were strong, reliable performers. (Author)

**21 July** 1962 and Stanier Black Five, No. 45253, arrives at Bath's Green Park Station with the 9.08 am Birmingham to Bournemouth passenger service. (Author)

to pollution. Sadly, the S&D won't be reborn to help fill the void and we are left to remember its glories and ponder what might have been.

Over the years, I have had a number of chance encounters that stimulated my interest in this railway and I slowly gathered many things to increase my knowledge and understanding of its history. Of greatest significance was a collection of S&D items that came up for sale at auction in 2009. These concerned Peter Pike, a Bath trained engineer, who became a fitter at Templecombe during the 1950s and 1960s. This collection included all his handwritten logbooks. Through these items, an intimate picture of all that happened at the sheds emerges in their last few years. This includes

**Towards the** end, 'specials' became the order of the day. Here one of the S&D's stalwart 7F engines pauses to pick up its passengers. (Author)

the variety of engines passing through his hands, the problems they faced in keeping them going in difficult circumstances and the minutiae of day to day life. When leafing through the pages of these well-thumbed, oil-stained books, one can almost smell the smoke and hear the clatter that accompanies heavy engineering. Theirs were, by any standards, hard but worthy lives and sat at the centre of all that this railway achieved. Men such as this served the engines as much as the footplate crew and ensured the line kept going and the company remained afloat as the end of steam approached and the line was due for closure.

The S&D's engines reflected the different phases of the railway's existence and the course of Britain's transport history. There were occasional diesel interlopers, but for the most part, steam dominated all workings from beginning to end. And what a variety of engines there were. For such a small line, the assortment included those built specifically for the company mixed with classes 'seconded' from other regions across Britain – the London, Midland and Scottish and the Southern Railway amongst them. It was this mixed bag that added to the S&D's distinctive character and made a summer Saturday overlooking the line such a wonderful spectacle to behold.

Having gathered material, and read many excellent books on this singular railway, it was a privilege to be asked to write a volume that focusses on the engines that worked this line throughout its life. My fascination with them and the line that carried these locomotives across the Mendips has not diminished with time. I hope this addition to the S&D's history captures the spirit and the variety of this magical railway.

# Chapter 1
# A LINE CREATED AND REMEMBERED

Throughout its life, this railway collected some interesting soubriquets – 'Slow and Dirty', 'Serene and Delightful' or simply the S&D. Whether flattering or disparaging, they seemed to sum up the character of this unique institution.

In some ways, it was a branch line that had ambitions to be much more but lacked the clout to do so. And yet in Great Britain it was probably one of the most demanding of routes, calling for immense skill and strength

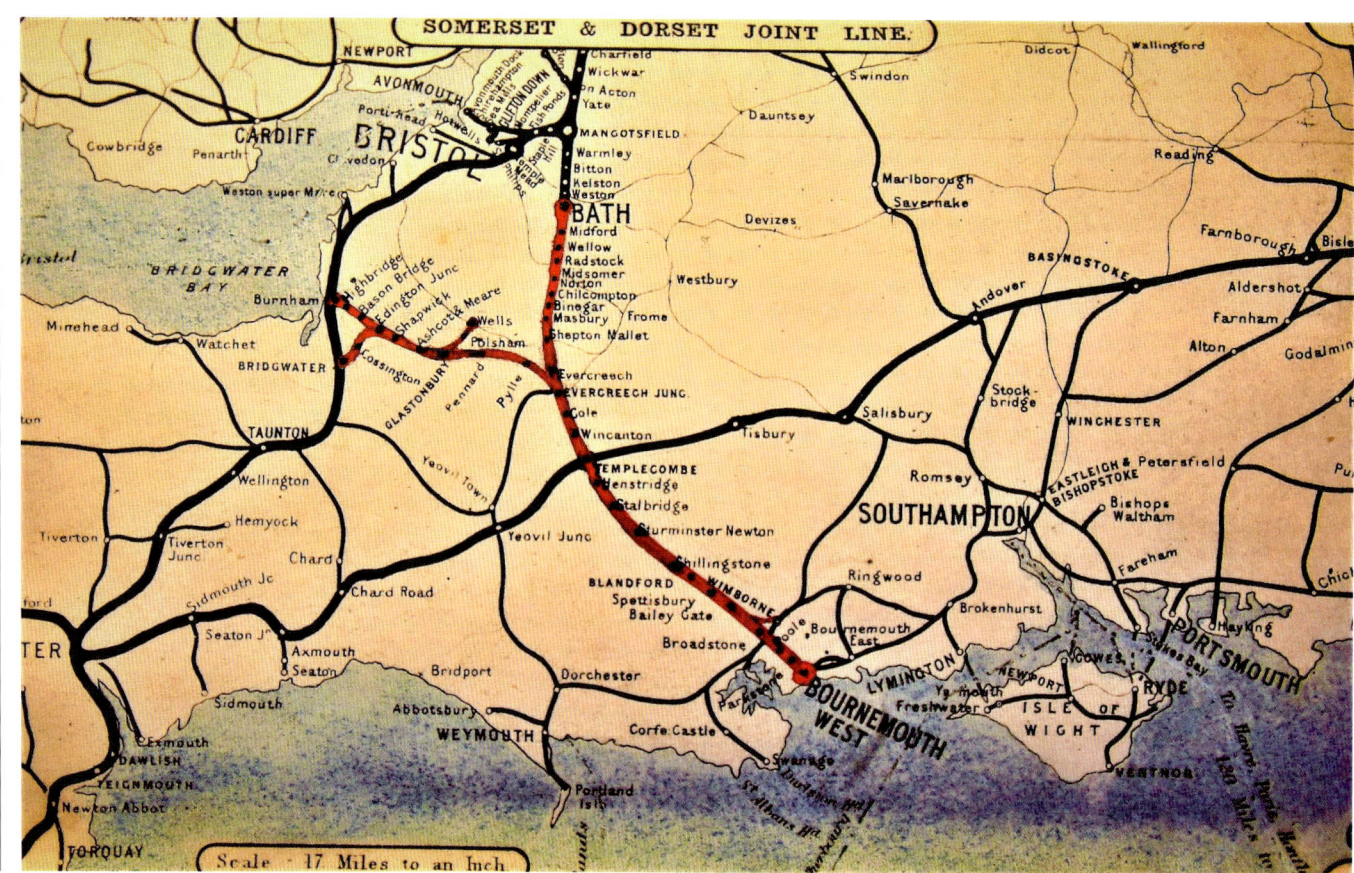

**The Somerset** and Dorset as it appeared in 1905 and recorded in one of the driver's notebooks. (Author)

from its footplate crew and its engines.

As a business, it retained a strong sense of individuality and this bred an innate, singular culture and way of doing things that reflected the countryside it traversed and the people who populated the line. A concentrated and distinctive community – Somerset and Dorset with a touch of Wiltshire thrown in, though it barely touched this county's boundary just south of Bath. And it remained so until closure in 1966, when more than a hundred years of acquired skills and dedication were swept away, seemingly without a second thought. In an act of cultural vandalism, many of its bridges were simply blown up, tracks lifted and the infrastructure dismantled. If BR's Western Region sought to eradicate all memory of this singular railway, and hoped there would be no resurrection, they succeeded up to a point, but its supporters are made of sterner stuff and have refused to let it die. BR's vandals have now gone, themselves the victims of modernisation, yet the image of the S&D is as strong now as it ever was. When choosing to destroy it, BR's managers should have noted the strength of those who built and ran this railway and their never-say-die attitude in the face of all challenges. From the first, these traits were only too apparent.

Britain's railway boom of the mid-nineteenth century, encouraged by the promise of this emerging technology and the quickly growing demands of the country's industry, soon attracted speculators. Some had sound business proposals, others were simply gamblers seeking a quick profit, so had a dubious appreciation of the need or the risks involved. Companies were formed, funds raised, schemes mapped out and gradually a lattice work of new lines spread across Britain in a way that often defied logic or rational plan. In this rush for expansion, two companies came into existence that would, in time, come together to form the S&D Joint Railway. The first of these was the Somerset Central, which

**The creation** and expansion of railway lines was underpinned by Parliament through a series of Acts. Here the key pieces of legislation leading to the formation of the S&D are carefully recorded on the inside cover of a company register. (Author)

received Royal Assent in June 1852 to construct a 7ft 0¼in broad gauge line between Highbridge and Glastonbury. It was an idea first mooted six or so years earlier, when a group calling themselves the Somersetshire Midland Railway had proposed a line from Bruton to Highbridge, where a maritime link to the Bristol Channel existed. The scheme came to nought, but a germ of the idea survived in the minds of the Somerset Central's management and received the backing of promoters at a meeting at the Railway Hotel in Bridgwater on 1 December 1851. The company's aim was a simple one – to provide an alternative to the Glastonbury Canal which had opened in 1834 to encourage trade and help improve drainage in an area prone to flooding. A new railway line, built in part along the canal's towpath, was thought the best way of re-invigorating the local economy. The opening of the line in 1854, offering six trains a day in each direction, sounded the death knell for the canal. However, it was hardly saturation coverage so could be managed by a single locomotive and a set of carriages, supplied and maintained by the Bristol and Exeter Railway company at Highbridge. One assumes that passenger trains were interspersed with some goods traffic, otherwise the aim of replacing canal barges would remain an unrealised ambition.

Whilst this line was being developed, the board of the recently formed Dorset Central Railway were preparing a rather more ambitious scheme. In this case they planned a 4ft 8½in standard gauge line that would go to Bath and from there north to the country's main business and industrial centres. With such a connection in place, it was believed that trade would increase and, with it, the area's prosperity. Such a project would be very costly and would, undoubtedly, have struggled to raise sufficient capital in the 1850s.

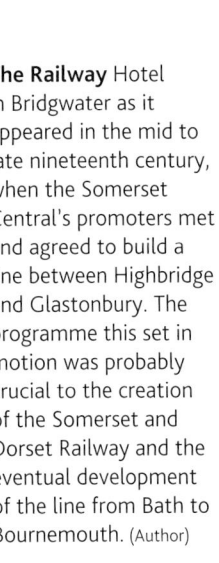

**The Railway** Hotel in Bridgwater as it appeared in the mid to late nineteenth century, when the Somerset Central's promoters met and agreed to build a line between Highbridge and Glastonbury. The programme this set in motion was probably crucial to the creation of the Somerset and Dorset Railway and the eventual development of the line from Bath to Bournemouth. (Author)

## A Line Created and Remembered • 15

By then, the railway mania that had gripped the country was dying away and investors were far less likely to gamble on any speculative venture. If any advance were to be made, projects such as this could only be tackled in a piecemeal way. To this end, approval to construct the first section from Wimborne, where it would link with the London and South Western Railway, to Blandford was sought in 1856 and construction was completed in 1860. Without its own locomotives or rolling stock, the Dorset Central handed responsibility for working the line to the LSWR. At this stage, this company was expanding their network and had, in 1847, taken on routes to Southampton and Dorchester, and planned to extend their influence into Devon and Cornwall. This made them a strong commercial presence in the area and one that was eager to exploit any new line, no matter how minor, to boost its business.

As will always be the case, once a business becomes established, plans for expansion soon follow. In both cases, the Somerset Central and Dorset Central soon put forward plans for extensions – Highbridge to Burnham, Glastonbury to Wells and Glastonbury to Cole and Bruton, whilst further south, a line from Blandford to Cole and Bruton was proposed. When completed, these lines would link the Bristol and English channels and bisect the Bristol and Exeter Railway Company's broad gauge line and the GWR's Paddington to Devon service. At the same time, it cut across the LSW's Waterloo to Exeter line at Templecombe. However, the question of incompatible gauges, and the need to transfer passengers and goods between them, led to a re-appraisal. Rather than convert all track to a single gauge, a compromise agreement saw a third rail provided which allowed Dorset Central trains to traverse the northern section of the line. This dual working would continue for many more years, as

**Glastonbury as** it appeared in the nineteenth century. This small market town became the primary focus of the Somerset Central Railway's ambitions. Linking it to Highbridge, where it intersected with the Bristol to Exeter broad gauge line, and then onwards to the coast, was thought to be the best way of re-invigorating the economy of the area. A limited aim that was soon subsumed by much grander ambitions. (Author)

**Templecombe late** in the nineteenth century. Although presenting a charming picture of rural life, it also sums up the problems faced by the S&D in trying to generate revenue in a sparsely populated area. Many places like this village were linked by the railway so generated little passenger traffic and only small amounts of goods. If the directors hoped that the railway would encourage more people and industry to the area, it was a hope destined to die stillborn. The link to Bath and the North was, in reality, a final throw of the dice. (Author)

a battle between the advocates of each gauge persisted. While this happened, and before the line was complete, the two founding companies considered their future and pushed through an arrangement which would see them work as one.

As the two companies formulated their plans and saw their lines slowly develop, some level of cooperation between them seemed mutually beneficial, especially in view of the costs involved. As Robin Atthill recorded in his 1967 book, this issue took on added weight in the autumn of 1859 when the Somerset Central's board discussed a subscription to the other company. This, it was believed, would confer 'mutual running powers over the whole of the two systems as soon as the extensions were completed'. Debate and discussions seemed to have run on without reaching any firm conclusions. Then, in February 1861, the Somerset Central's directors, in an effort to resolve the matter, proposed that both companies should work as one and bear responsibility for day to day operations. This proposal was approved at a special general meeting held four months later on 29 June. From this point, the plan gathered pace and in August it was confirmed that 'amalgamation should take place at a future time, so as to secure entire unity of action and economy in management'. Whether it was a fait accompli, bearing in mind the business and funding pressures both companies faced, or a considered expansion plan is largely immaterial; the die was cast. On 9 May the following year, during a second special meeting, a Bill of Amalgamation was finally approved and received Royal Assent three months later.

The new organisation was probably launched in a spirit of hope, though such optimism would soon have been quelled by a cold, hard look at the company's finances. Over-capitalisation was endemic, with the value of shares and debentures well in excess of fixed asset values. Receipts, whilst having the potential to rise once the line was complete, were deemed insufficient to meet future capital expenditure, let alone existing costs. Their dilemma was only too clear. Press on and hope for the best, giving life to the business mantra 'speculate to accumulate' or stand still and hope that sufficient traffic could be generated to make the business a going concern. In reality, neither option made good business sense in the light of their parlous financial state, but the commitment had been made and the more optimistic souls hoped that the line's completion would help to attract more customers and increase revenue.

Over the next fourteen years of independent existence, the struggle

**The coming** together of the Somerset Central and Dorset Central in 1862 was secured with this seal. The new company would last for 13 years and for most of this time would struggle to establish itself and remain solvent. (Author)

**It is** very easy to focus on the activities of the directors and speculators when the history of a new line is being described. Then, once it is in place, the business of day to day running attracts most attention, as do locomotives and rolling stock. The missing link and the key to the existence of any railway – the men who struggled in the most appalling conditions to build it – is largely forgotten. Their work was unregulated, hard, dangerous and mostly manual. This was none more so than on the Somerset and Dorset, particularly over its northern section to Bath. A considerable number of labourers, as pictured here, came from Ireland, where poverty and famine were rife, to find work and many paid with their lives for the privilege. (Author)

**An interesting** illustration showing the route of the Bath extension in advance of its construction. Of particular note is the title, which strongly implies some sort of proprietary interest in the S&D, by the Midland Railway and the LSWR, even though it was an independent line. If so, it hints at the shape of things to come. (D. Neal)

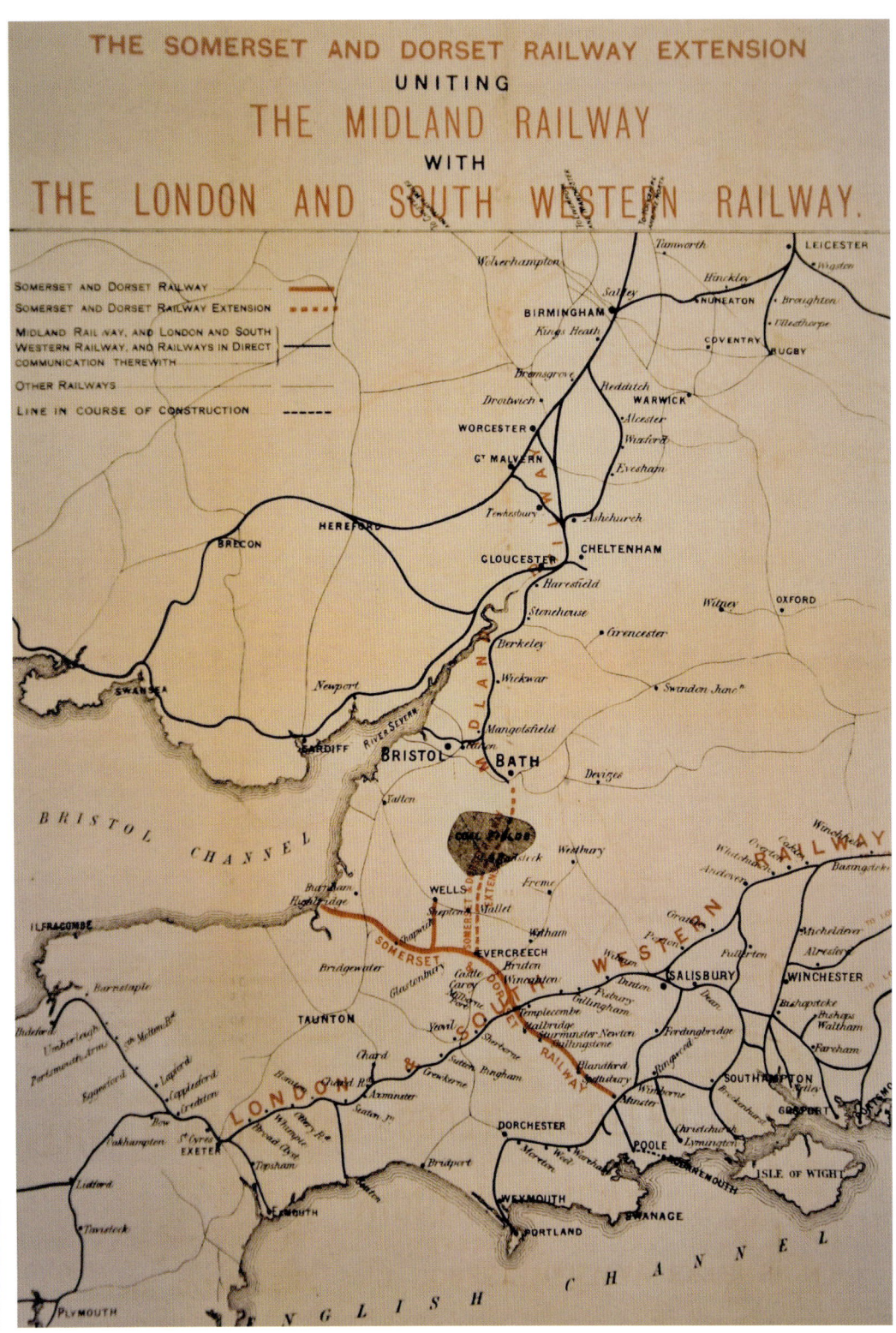

to make the line pay continued without let up. Once again, Robin Atthill captured the essence of the battle they faced. He surmised that there were two phases in this campaign:

'Nine years (from 1862 to 1871) of hectic planning and frustration, including four years of virtual bankruptcy while the company's affairs were in the hands of the official receiver; and a further five years (from 1871 to 1875) full of equally frenzied activity, which nevertheless resulted in the fulfilment of the final scheme for the extension of the system, even at the price of complete financial prostration.'

The new company, once formed, pressed on with the development of the line between Templecombe and Blandford, which opened on 31 August 1863. Yet, despite the problems faced, they didn't allow their financial difficulties to cloud their ambitions, which still centred on making the connection to Bath and Bristol and so open up the lucrative trade route to the north. But first there were nine years of financial stringency to navigate successfully, with the constant risk of collapse ever present. The main problem was that the links to the two Channels did not attract the trade hoped for, so it fell to the area's main industry – agriculture – to provide business, with passenger traffic providing a poor source of revenue. In such a sparsely inhabited area, it couldn't be otherwise. The two cities to the north might provide the trade they needed, with the potential for business with North Somerset coalfields providing an added spur to this plan. Nevertheless, it was an ambition likely to remain unfulfilled whilst the company remained in the fiscal doldrums,

**One of** the major civil engineering projects on the Bath extension was Tucking Mill Viaduct, here undergoing modification to allow a doubling of the line in the early 1890s. The Johnson 0-6-0 tender engine, No. 35, is captured heading south towards Wellow having passed through Devonshire and Combe Down tunnels and approaches Midford Viaduct. (Author)

**Reconstruction and widening work continues here with the aid of a narrow gauge line laid to bring building materials to the site, with a temporary platform provided to aid this work. Information about the engine is not written on the back of this old, sepia print, but the location is recorded as 'Charlton Viaduct near Shepton Mallet in 1888'.** (Author)

unable to manage its debts and raise fresh capital.

Post-1870, this changed. Having been placed in administration, the company was managed by receivers appointed by the Court of Chancery. Their application of stringent economies brought a degree of control to the business and improved its financial position considerably. This allowed them to begin raising funds again through debentures up to a total of £796,950 against an estimated asset value and cash total of £1,324,165. The books may have been nearing a reasonable balance, but this didn't mean the company was truly solvent or profitable. Nevertheless, in a spirit of optimism, perhaps ill-placed, the Bath extension became an active project, aided in part by the arrival of the Midland Railway in the city on 4 August 1869. Although only a branch line from the Bristol-Birmingham main line for the company, it opened up the possibility of the much-desired northern link for the S&D if they could reach Bath.

An Act of Parliament in 1871 authorised the extension and allowed the company to raise £480,000 in shares, mortgages and debentures; just sufficient to fund all aspects of the project. Work began a year later and on 20 July 1874, the first revenue-earning train left Bath. Despite the comparatively short construction period, it proved to be a gruelling challenge, only made possible by the employment of 3,000 or so labourers working round the clock, often in appalling conditions.

The route chosen to Bath, taking in Shepton Mallet, Binegar, Chilcompton, Midsomer Norton, Radstock and Midford, had many difficult obstacles to overcome. These included the Mendip Hills, which reached a peak near Masbury of 811 feet, and the hilly, valley strewn and very difficult approach to Bath itself, where extensive tunnelling and bridgework were necessary. Despite these challenges, they pressed on, occasionally suffering cash flow problems which led to disputes with its main contractor, T. & C. Walkers, over unpaid bills. These were overcome by the transfer of shares

**With the** frontage of a major terminus, one might be forgiven for thinking that this station in Bath was the seat of a major company and served an area of some affluence and importance. Certainly, the Midland Railway could make such a claim, but the S&D could not aspire to such a status. Nevertheless, as captured here in the late nineteenth century, this distinctive station had style and class and easily rivalled or exceeded the GWR's Bath Spa station a mile or so away. (Author)

**Robert Arthur** Read, a key figure in the history of the S&D, as he appeared in the mid-1870s, when his hard work on behalf of the company as Secretary, then Managing Director, resulted in his being appointed General Manager in 1874. His connection with the Somerset Central began in 1853 and he remained as an active participant in the work of the line until 1891. (D. Neal)

and debentures to Walkers, with the residue made up of Lloyds Bonds and, when available, cash. Undoubtedly, the efforts of the company secretary, Robert Arthur Read, and the Consulting Engineer, W.H. Barlow, were crucial to the eventual completion of this project. But from the beginning, Read had played an important part in the history of the S&D. He was a constant feature of each stage of development, growing in experience himself as the line gradually expanded. In any history of the company, the part he played must feature large. This was especially so from 1874, when the link to Bath finally opened, and he was appointed General Manager. His presence was and would remain a dominating one.

Robert was born in 1830, the third child of Stroud schoolteacher Paul Read and his wife Mary. On leaving school, he found work as an administrator and became Clerk to the Bailiff for the town of his birth. Clearly an ambitious young man, he then found employment with the Bristol and Exeter Railway, where he worked for the Company Secretary. When the first Secretary of the Somerset

Central joined the South Eastern Railway, Read was recruited as his replacement at a salary of £200 per annum. There then followed a move to Glastonbury, where he would remain until the late 1880s. Here he and his wife, Mary, raised their ten children, while, in the background his career went from strength to strength. As an administrator, he was clearly gifted and led the company's directors through each stage of development, but he was more than this. From an early age he proved to be an able businessman, accumulating some personal wealth in the process and various directorships, one being the Poole and Bournemouth Railway. These two aspects of his personality would prove essential to the company as it gradually grew. He, it seems, brought some reality to the board's plans, though even he couldn't curb all their ambitious and, perhaps, unrealistic schemes. If anyone could justly take credit for the company's survival and the avoidance of financial catastrophe it was he. But there was only so much he could do with a railway serving a community with a small population and little industry. His value to the company was probably best summed up in 1862 when the directors rewarded him for steering the business 'through a period of great financial difficulty…under the circumstances they can only record their unqualified approbation of the manner in which Mr.Read has discharged his duties'.

His professionalism and insight remained a driving force in the struggles ahead, supported by the Engineer in Chief, Charles Gregory, and the Locomotive Engineer, Robert Andrews, who was succeeded by Frederick Slessor in 1868 and Benjamin Fisher in 1874. Between them, and with the approval of the Board, of which he was Managing Director from 1871 to 1874, key issues such as locomotives and rolling stock were discussed and agreed. To Read would have fallen the difficult task of generating funds to meet these requirements and, it seems, occasionally using his own money to help the company continue operating. He did all this with a personal tragedy unfolding in the background. He was widowed at the age of 50 and was left to bear sole responsibility for seven of their children who still resided at home. His daughter, Annie, took on the role of housekeeper until Robert re-married, became a father again, and moved the family from Somerset to a substantial house in Surbiton Hill Road near Kingston, Surrey, supported by five servants, where his active connection with the S&D continued until 1891. By this stage, he was a Lloyds Underwriter and had also become a Director of another City based company, so was a well-established and successful figure.

The link with Lloyds may account for the success the S&D displayed in staying afloat despite its many problems. Armed with many City connections, doors could be opened, and new sources of funding tapped. Attracting investors would have been part and parcel of Read's ethos and reflected his wide business interests. It was a role that would continue as the company faced its next significant challenge and with it another battle for survival. Overstretched by its ambitious plans, it could only survive if it established some sort of partnership with another company, with the Midland Railway, GWR, Bristol and Exeter and the LSWR all being possibilities. With the line to Bath now complete, the S&D was a more attractive proposition than it had once been, but the national debate over which gauge should become the British standard put it in the middle of a much bigger battle taking place between supporters of each system.

One or the other seemed likely to come out on top but in the 1870s it was far from clear which one this would be. In the meantime, the cost of the Bath extension, and lack of revenue to service its debts, meant that Read and Co had run out of options and amalgamation was unavoidable. But with whom and on what terms?

By this stage, the Bristol and Exeter Board had probably seen the writing on the wall when it came to broad gauge. As a result, its directors proposed adding a third rail to its entire system so that it could link with other networks. Meanwhile, the GWR remained committed to broad gauge, whilst the Midland and the LSWR, amongst others, championed the narrower gauge. It was from these companies that a partnership was sought and in May 1875, Read approached the GWR, which duly announced itself ready to take over the line, subject to terms and conditions being agreed. The negotiation was not a quick one and by 12 August, progress had ground to a halt. In the interim, GWR managers contacted Archibald Scott, the LSWR's General Manager, in a bid to spread the risk and the costs. They proposed that the

# A Line Created and Remembered • 23

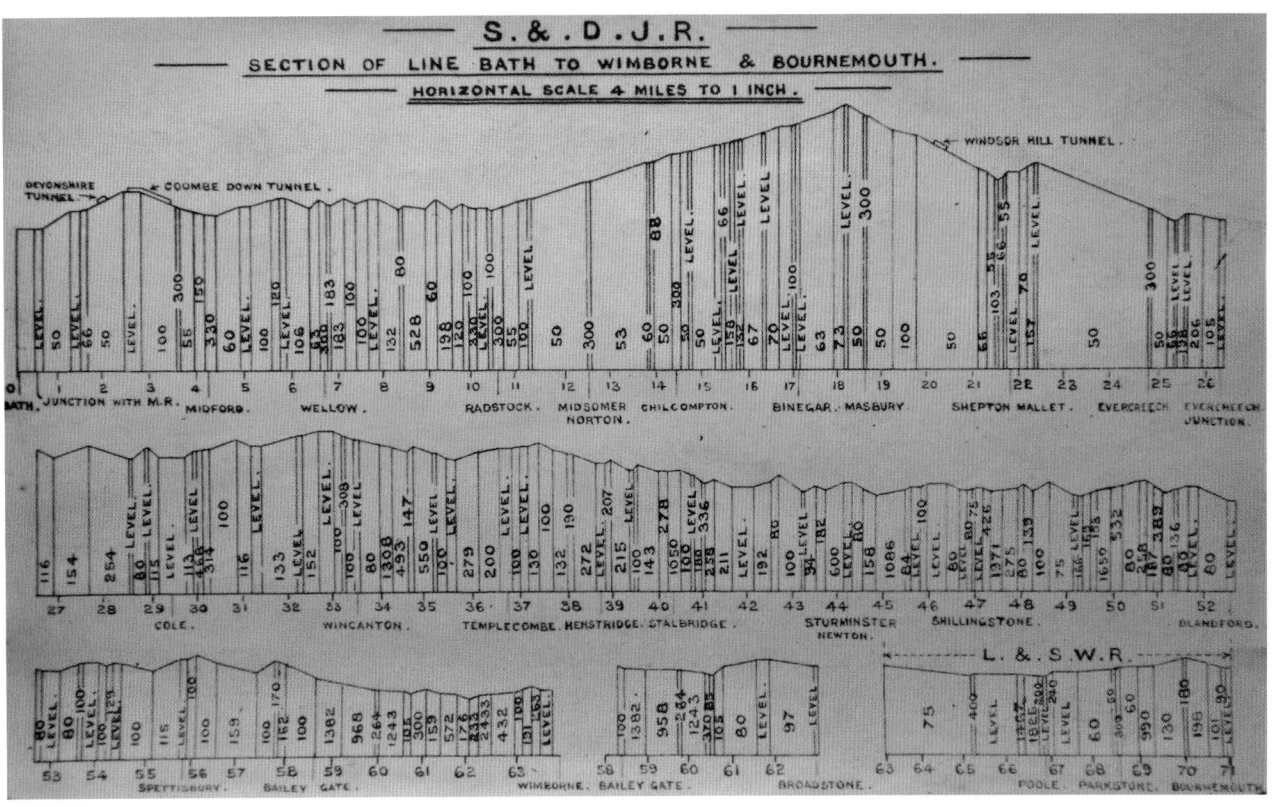

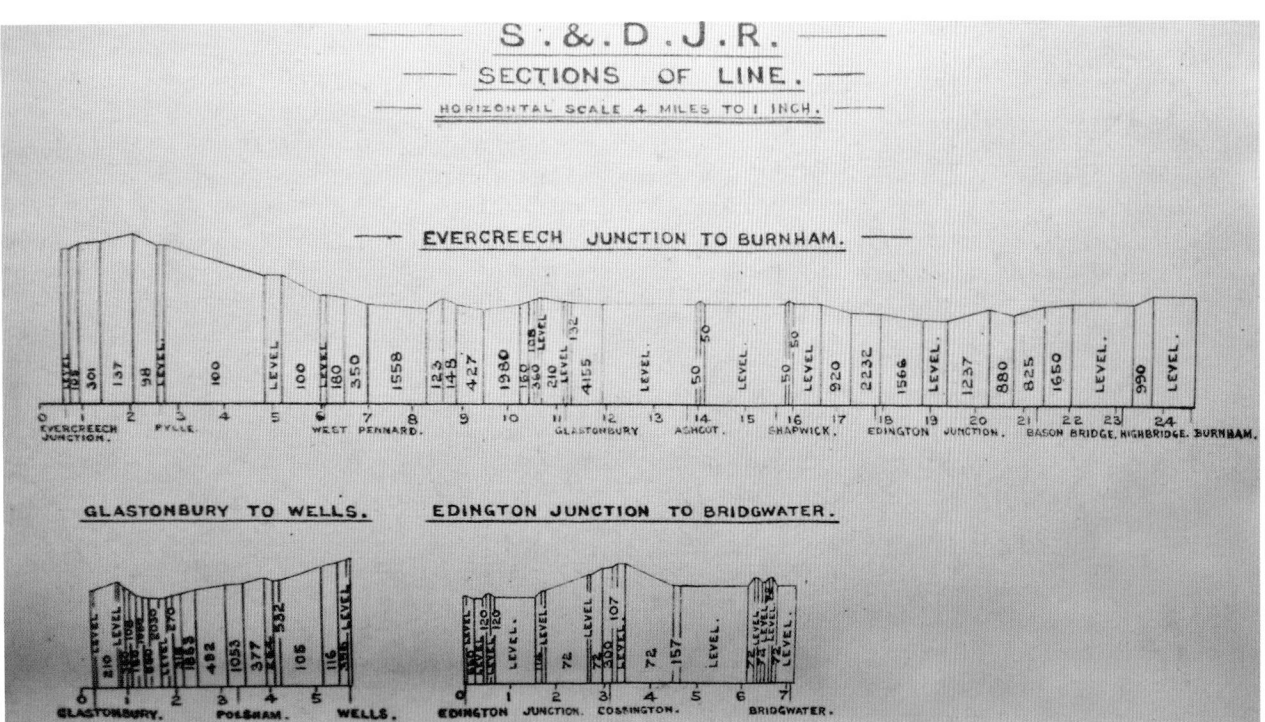

**The S&D** as it appeared towards the end of the nineteenth century in relief form when leased to the Midland Railway and the LSWR. Mile for mile, it probably had as many challenging features as any line in Britain. (Author)

**The S&D** was a most picturesque railway and fitted well into the landscape through which it passed, as demonstrated here by the line as it wended its way through Radstock with its sweeping curves following the contours of the land. (Author)

**The company's** new coat of arms. (Author)

Waterloo based company operate the southern part of the railway, leaving the rest to the GWR and the Bristol and Exeter Railway. Scott is reported to have been deeply troubled by the S&D's proposal to 'sell out' to the GWR, presumably because they were their chief rival. He immediately contacted James Allport, the Midland's General Manager, to seek a more favourable arrangement and together they formulated a joint leasing agreement. Their approach to Read was successful and he recommended acceptance to his board, which they duly approved on 20 August. Reaching any other conclusion was most unlikely in the circumstances, with the 'takeover' more likely to provoke feelings of relief rather than regret.

This wasn't the end of the matter, though. The GWR, under the chairmanship of Daniel Gooch, sought legal redress claiming breach of faith by Scott and the company. They then went on to accuse the Midland of reneging on an 1863 agreement in which they had promised not to extend their services south beyond Bath. The legal wrangle ran on for some time but to no avail and the joint MR/LSWR leasing arrangement received Parliamentary approval on 13 July 1876. A little over two weeks later, perhaps hoping to strengthen its broad gauge credentials and ensure that the Bristol and Exeter didn't opt for the smaller gauge in its entirety, the two companies joined together. Against such odds, they were unlikely to succeed in the longer term. But Brunel's spirit was difficult to cast off and it would take another sixteen years to complete the changeover, with the last broad gauge service being run from Paddington on 20 May 1892.

For his part in rescuing the S&D, Read was awarded a sum of £5,000 by the directors. At the same time, he was selected to be General Manager and Secretary of the Somerset and Dorset Joint Railway as it now became. Clearly this wouldn't have happened if either new partner had harboured any misgivings about him. Under his guiding hand, the railway would enter a period of comparative calm, propped up by substantial amounts of rent paid each year by the leaseholders. In the first year, these totalled £43,056, rising to £57,408 three years later, plus one tenth of any receipts over £114,816 per annum. This enabled the railway to stabilise its finances and raise more share capital, with the risk being shared between all three companies. However, the two new partners were probably left in little doubt about the level of risk they had taken on and the problems they seemed likely to face. It was a point pressed home only a few weeks after their leases began.

A serious rail crash at Radstock on 7 August 1876 killed thirteen and injured another thirty-four people. The investigation revealed the parlous state of the railway caused by insufficient funding, an inadequate infrastructure and poor management at all levels. Highly critical press coverage, followed by a scathing official report, which detailed all the S&D's failings, couldn't be ignored. However, this was a time when railway safety carried with it very little, if any, statutory legislation, leaving companies to operate as they saw fit. With profit and loss being the main driving force, corners were inevitably cut. As a result, the number of crashes and casualties soared until the government was forced to grasp this unpleasant nettle and begin to control a largely unregulated industry.

**The Radstock** rail disaster as portrayed in the popular press. Lithographs were then a common way of capturing such events and gave them a dramatic edge often found missing in photographs of the time. Nevertheless, it was a shocking incident that might have killed off the S&D if a new leasing agreement hadn't been sanctioned by the government then ratified only a few weeks earlier. (D. Neal)

**As the** new century approached, the S&DJR's infrastructure was well-established and included a number of sheds as well as a plethora of stations. At the northern end of the line were the facilities at Bath, which by the 1890s included separate sheds for Midland and S&D engines and a Fitting Shop (as captured in this late century drawing). (Author)

The trauma of this accident gradually died away and the line settled into a more stable period of operation, though the line's many deficiencies could not be resolved overnight and would take many years to overcome. But at least it now had the structure and support it needed to make improvements and bring it in line with the industry generally. However, this was hardly a huge advance, especially in a business that could still do largely as it wished without regulatory interference and independent scrutiny. This would change following the Armagh rail disaster on 12 June 1889, where poor practice and incompetence killed 80 and injured another 260. It is generally agreed that modern day safety and operating standards had their true genesis in the aftermath of this dreadful accident. But it would be a slow process to completely instil safety first principles as many more accidents and deaths would graphically bear witness.

Nevertheless, over the coming decades the S&DJR cemented its position and continued to develop its infrastructure, particularly stations and workshops. This was made possible by Read's leadership, backed by the spreading influence of the Midland Railway and the LSWR, but he could not do this alone and it was here that Robert Armstrong Dykes came to prominence.

He was born to Henry and Jane (nee Armstrong) Dykes in Bristol on 9 October 1834. On leaving school, he initially found employment as a clerk with the Bristol and Exeter Railway, where his father was Traffic Superintendent. From here he moved to become a railway inspector with the Midland Railway. He then transferred to Spalding in Lincolnshire to become

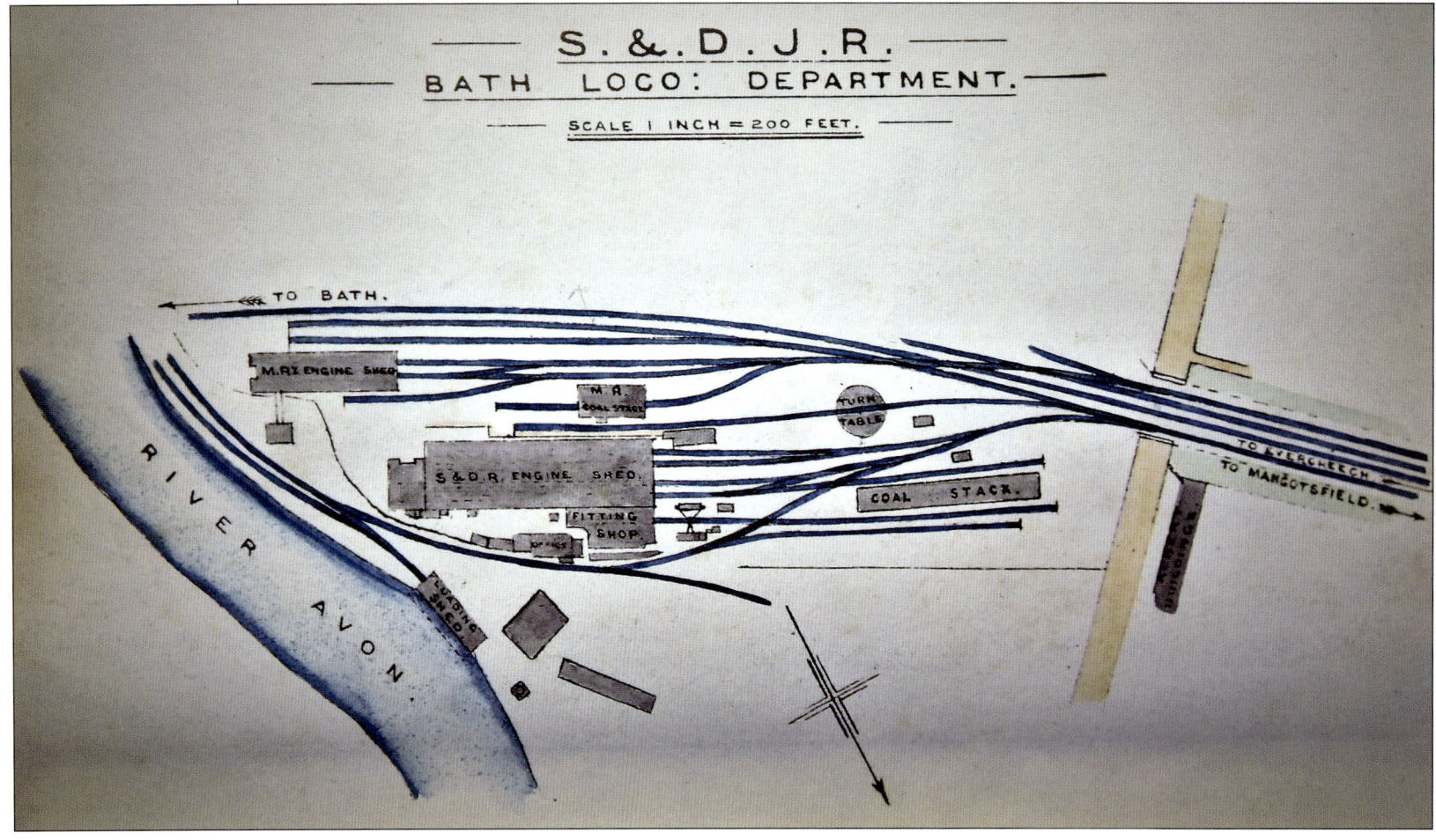

Traffic Superintendent of the Great Northern and Midland Joint Railway, as his Census return for 1871 confirms. By this time, he was married with children and well established in his career and it was here that the S&DJR, on the lookout for a new Traffic Superintendent, found and recruited him in 1876. With both he and his wife Mary having strong links with Bristol, the move must have been agreeable to them both. From here, a sixteen year mutually beneficial association began, during which Dykes would turn his considerable skills towards the line's continuing development. By the time Dykes had taken up his post, the S&DJR's directors had decided to move its headquarters in Glastonbury, where it had been since the days of the Somerset Central, to Bath. Dykes and his family moved into 3 Norfolk Crescent with its views across the River Avon to the station and sheds. From this, one can assume that he was determined to be on the spot, so to speak, and closely manage his department. If so, he proved

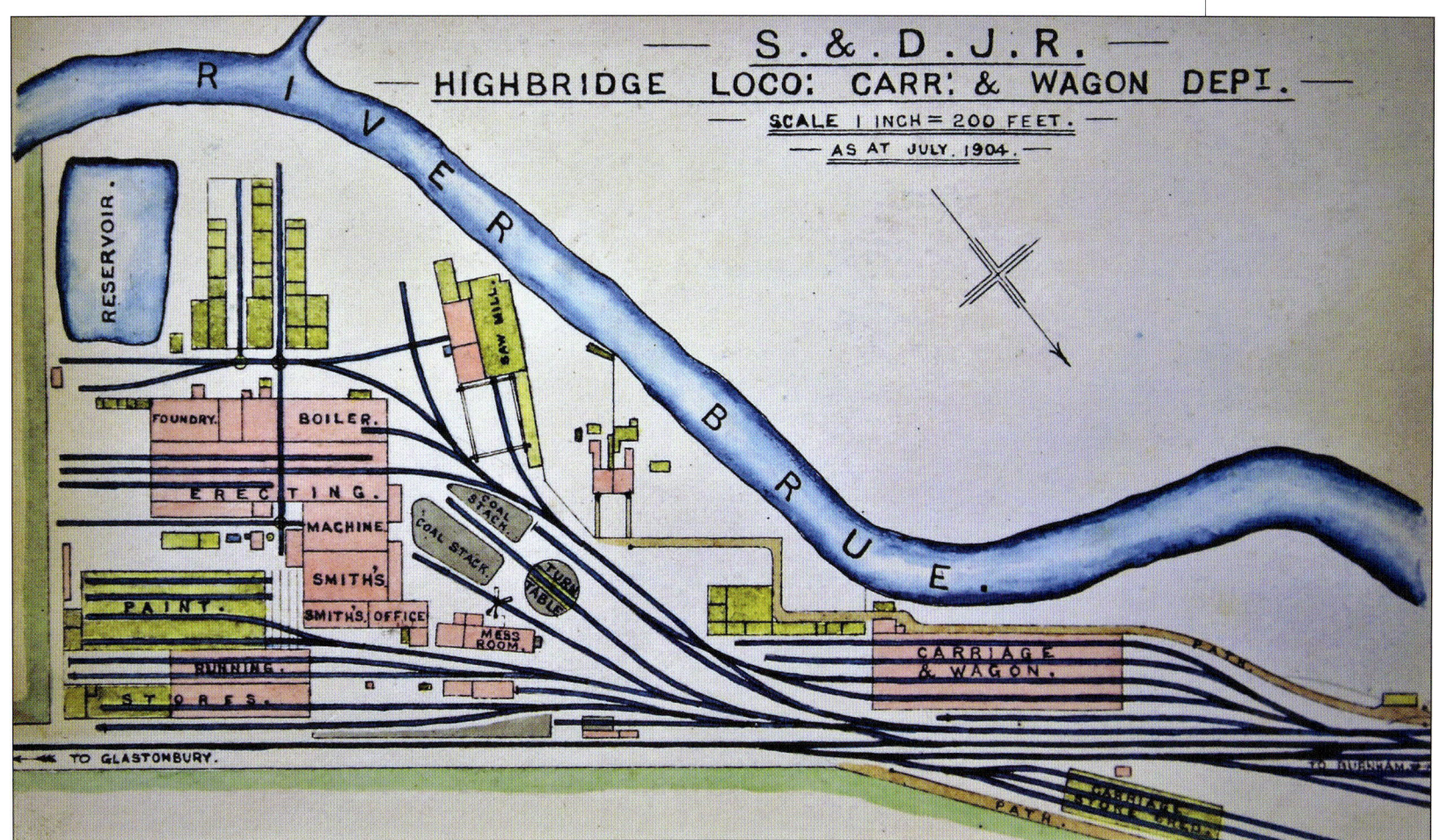

*Above and overleaf:* **From the** earliest days of the S&D, there had been a workshop facility at Highbridge for the repair of locomotives and carriages. As the company grew, so did this facility in the town. Extra land was purchased, and a complete range of workshops were constructed, including Erecting, Machine, Paint, Boiler and Spring Shops. In addition, there was a Drawing Office, sawmill and foundry. The primary purpose was the repair and rebuilding of engines and rolling stock, but the establishment also had the ability to construct new locomotives if required. This in fact only happened on one occasion and involved a tank engine for the Radstock colliery branches (an 0-4-2T in 1885). However, two others appear to have been assembled there for the colliery (both 0-4-0Ts in 1895). The drawing above shows the extent of these workshops, while the photographs bear witness to work carried on in the shops and the equipment used. Of note is the order and cleanliness of the workshops, surprising considering the nature of the work. All these pictures date from the period when Alfred Whitaker was Locomotive Superintendent. (Author)

ERECTING SHOP. HIGHBRIDGE.

FITTING SHOP ENGINE. HIGHBRIDGE.
CYLINDER 14" x 30". REVOLUTIONS 110 PER MINUTE. STEAM PRESSURE

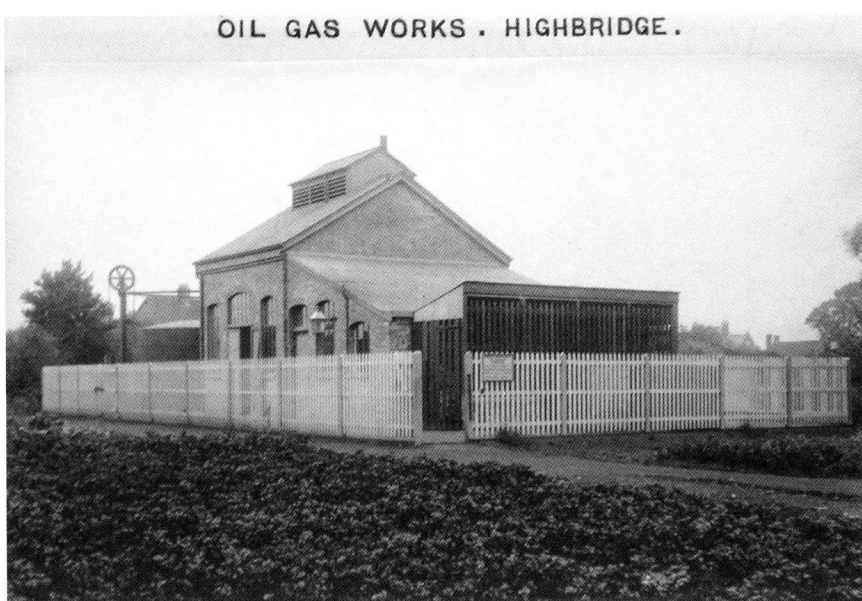

OIL GAS WORKS. HIGHBRIDGE.

STEAM & HYDRAULIC CRANE. HIGHBRIDGE WHARF.
TO LIFT 30 CWT. 2 STEAM LIFTING CYLINDERS 16 INS DIAM.
I HYDRAULIC LOWERING CYLINDER 7½ INS DIAM.

remarkably successful and exerted a positive influence on the way the railway operated.

In Dykes' experienced hands, a clear plan of action was developed which enabled the company to tackle its many infrastructure problems. Towards the end of his career, in an article in the *Railway Magazine*, he touched on the problems faced and the effort it had taken to turn things round. He recorded that:

'The fortune of the old Somerset and Dorset was certainly very varied, and not particularly prosperous. Various schemes were contemplated whereby to retrieve their failing fortunes …

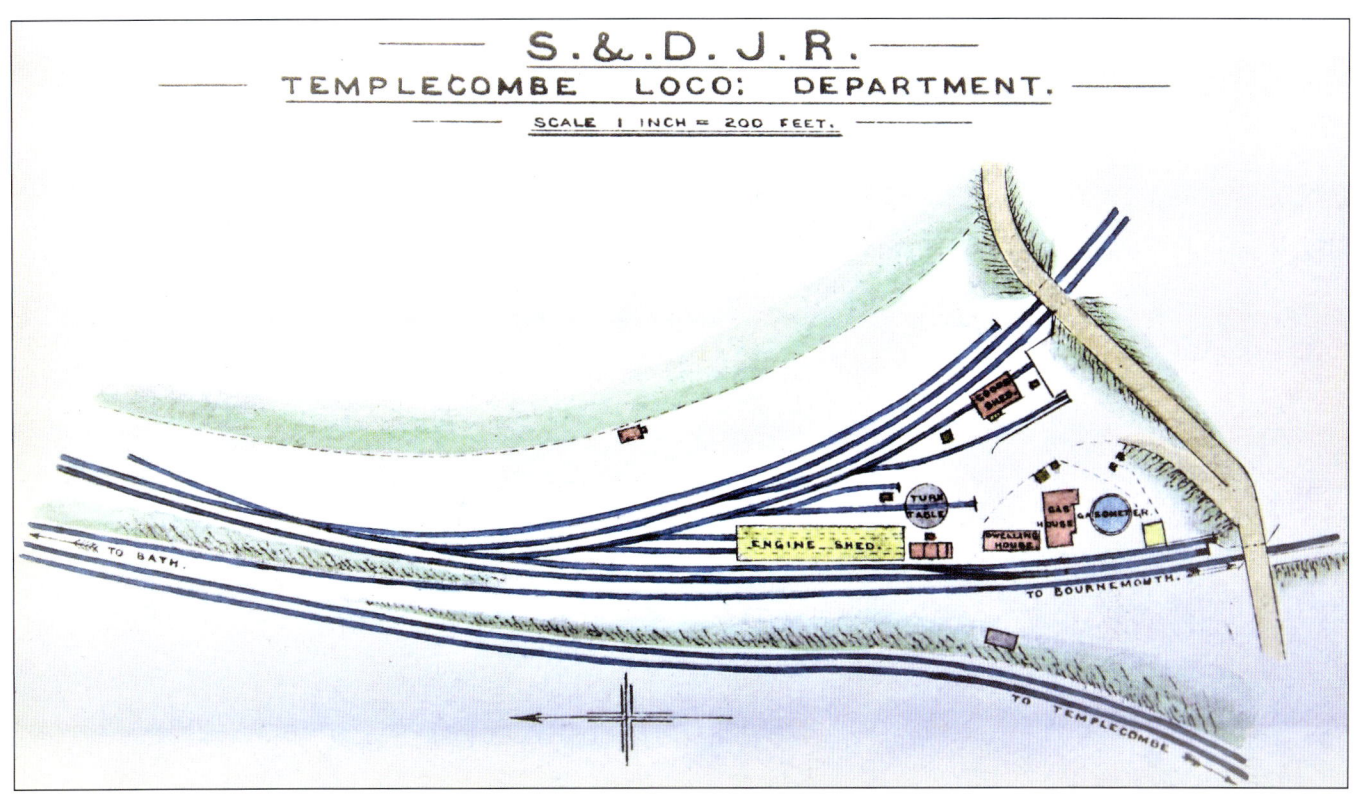

**Where the** S&DJR crossed the LSWR at Templecombe, there existed a shed and some support functions. Although much smaller than either Highbridge or Bath, fitters were numbered amongst staff permanently assigned to the establishment and carried out essential repair work in the shed or when locomotives broke down nearby. This, it would seem, reduced the amount of time lost if a locomotive had to be towed to Highbridge or Bath for repair. (Author)

**The description** on the back of this turn of the century print records that the picture depicts 'the shunting yard at Bridgwater and small engine shed' though doesn't record the engine's number. (Author)

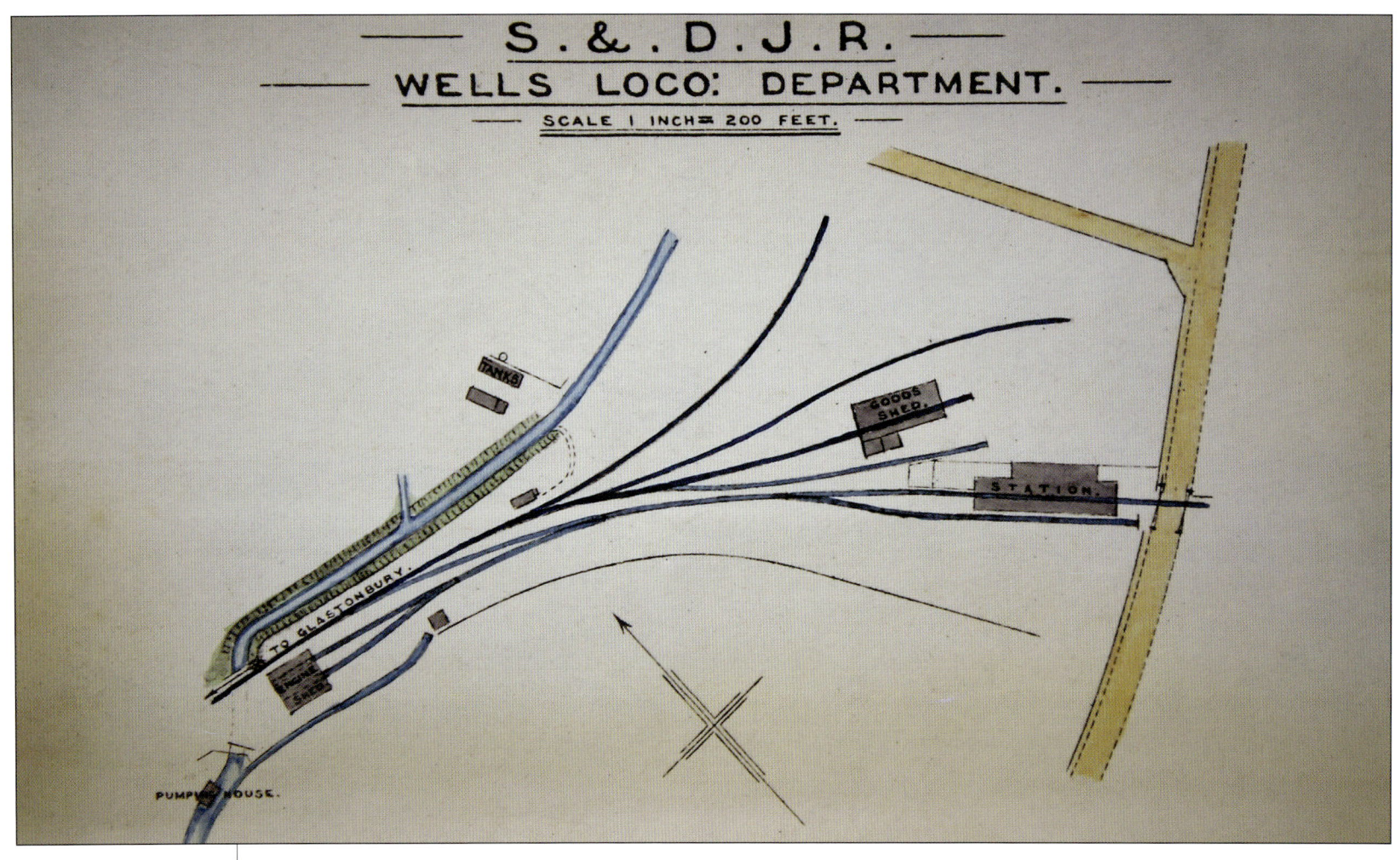

**Away from** the larger centres of activity, other stations had their own sheds and allocated engines, as demonstrated here at Wells. (Author)

it was not until the projected extension from Evercreech to Bath to join the Midland Railway was put forward that the old company found themselves able to improve their position'.

He then added, when questioned about the continuing interest in the line by the GWR and any influence Swindon might still exert, that:

'The last portion of the broad gauge was taken out about 1870, (although) the Great Western have the right to run their trains through Wells Yard only; a distance of nine chains.

'For some years I found I had a very uphill task before me, but by dint of unceasing labour on the part of Mr. Whitaker, Resident Locomotive Superintendent from 1889 [up to then it had been Benjamin Fisher and then William Henry French who presumably played some part in this work], Mr. Colson, Resident Engineer from 1876 and myself, and well backed by the owning companies, we succeeded in placing our line on a par with any in the Kingdom.'

There was little left untouched by these efforts. Locomotives and rolling stock numbers were increased, track was improved and doubled in places, a tablet system was installed, the workshops at Highbridge were rebuilt and extended and then there was work undertaken on signalling equipment, telegraph communications, staff recruitment, training and much more. None of

Templecombe Station.

**Templecombe Station** at the turn of the century on a fairly busy day. This is a scene so typical of railways at the time. With labour cheap, a large number of people were employed at most stations to ensure the smooth passage of customers and keep the infrastructure in good order. It was a number that would eventually be hard to sustain as wages increased. (Author)

this was achieved overnight, but by stealth over a number of years as funds allowed. Nevertheless, the Radstock accident probably ensured that safety was given the highest priority in all this work.

In mentioning Alfred Whitaker, Dykes introduced another figure whose work would prove of the greatest importance to the company.

Born on 22 July 1846 in Derbyshire General Hospital, to which his father Samuel was solicitor, he left school in October 1860 to join the Midland Railway as a clerk working for Matthew Kirtley, who was then Locomotive Superintendent. His employment records then show that in 1864 he was working in the Fitting Shops at Derby,

suggesting some sort of engineering apprenticeship was underway. Here he remained until 1869, when appointed to become Locomotive Foreman at Lancaster. Although only 23, his appointment suggests that Kirtley and his senior managers had identified some potential in the young man which they wished to exploit. This was followed two

**Three of** the key figures in the revival of the S&D's fortunes. (Left) Robert Armstrong Dykes, Traffic Superintendent. (Middle) Alfred Whittaker, the S&DJR's influential Locomotive Superintendent between 1889 and 1911. (Right) Alfred Colson the Resident Engineer and son of Thomas Colson who had been Engineer of the Croydon Canal. He began his career, with his brothers, as a railway contractor and was involved in the construction of the Somerset Central Railway. In due course he was employed by the company and became Resident Engineer in 1876. (Author)

years later by a level transfer to Bradford before returning to Derby in 1872 to become Assistant District Inspector. Several more postings followed, until, in 1889, he was chosen to become the S&DJR's Locomotive Superintendent based at Highbridge. The vacancy had suddenly opened up when William French resigned.

Coinciding with Whitaker's appointment, there was a fundamental change in the way the company and its partnership with the Midland and LSWR would operate. It is hard to say what gave rise to this transformation, but it probably centred on some dissatisfaction with the management and financing arrangements. The end result was two Midland Railway Acts in 1889 and 1891 which offered S&D investors the option of exchanging their holdings for Midland stock. Perhaps with memories of past financial problems still in the collective memory, many chose to take this route. Very soon, £2m out of £2.3m stock was exchanged on very favourable terms and passed to the two leasing companies. This was hardly surprising, considering the size and economic clout of the S&D's partners. But it meant that they now controlled the line more directly and soon moved to disband the long-established S&DJR board of directors and bring in their own men. In this case, an equal number from the two companies, which met for the first time at Waterloo on 3 February 1891. After so many years of active service, there was no place for Robert Read. But at 60, with other business interests to absorb him, it probably wasn't a pressing issue. He would live for another seventeen years, dying in Carshalton during 1908.

It was probably no surprise that Whitaker, a Midland Railway employee, was appointed at this time, especially as it had been agreed that this company would be responsible for locomotive policy across the S&D. Over the next twenty-two years, this influential man would guide the day to day operations of the line, linking with Dykes to ensure it ran as effectively as possible. One thing was certain, though, it was no longer an independent line but only one part of a network. It retained a distinctive character which didn't change appreciably until the end in the 1960s, but it would never enjoy any autonomy but be blown by the winds of change in the railway industry. This would be only too apparent in the locomotives and rolling stock that appeared crossing the Mendips over the next seventy or so years. The engines are the subject of this book and the later chapters detail their histories, but it is interesting to consider the rolling stock that graced the line as the new century dawned and, once again, we have records that Whitaker kept to inform our knowledge.

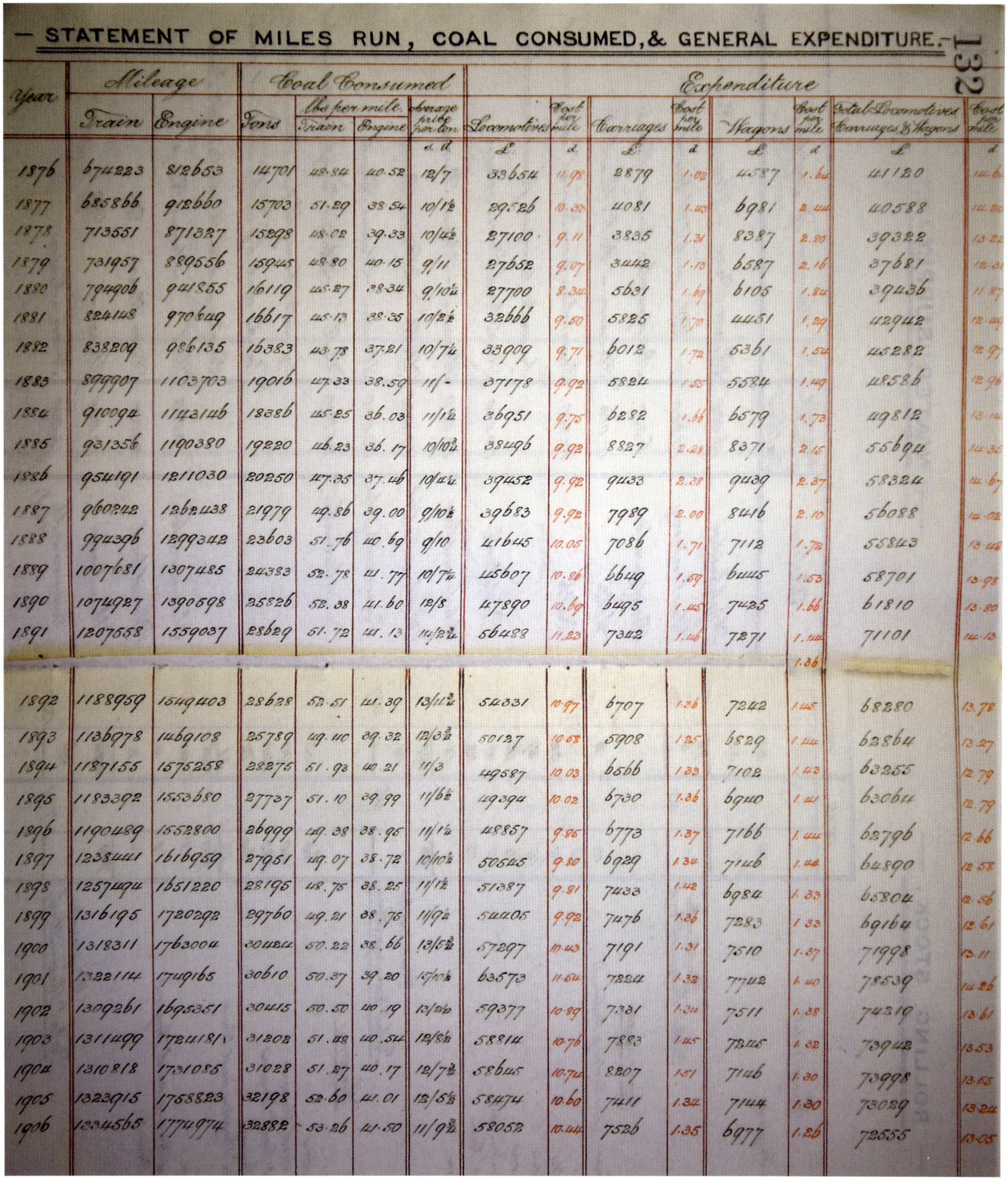

— STATEMENT OF MILES RUN, COAL CONSUMED, & GENERAL EXPENDITURE.— 132

| Year | Mileage | | Coal Consumed | | | | Expenditure | | | | | | | |
|---|---|---|---|---|---|---|---|---|---|---|---|---|---|---|
| | Train | Engine | Tons | lbs per mile. Train | lbs per mile. Engine | Average price per ton | Locomotives | Cost per mile | Carriages | Cost per mile | Wagons | Cost per mile | Total Locomotives Carriages & Wagons | Cost per mile |
| | | | | | | s. d. | £ | d | £ | d | £ | d | £ | d |
| 1876 | 674223 | 812653 | 14701 | 48.84 | 40.52 | 12/7 | 33654 | 11.98 | 2879 | 1.02 | 4587 | 1.64 | 41120 | 14.62 |
| 1877 | 685866 | 912660 | 15703 | 51.29 | 38.54 | 10/1½ | 29526 | 10.33 | 4081 | 1.43 | 6981 | 2.44 | 40588 | 14.20 |
| 1878 | 713551 | 871327 | 15298 | 48.02 | 39.33 | 10/4½ | 27100 | 9.11 | 3835 | 1.31 | 8387 | 2.80 | 39322 | 13.22 |
| 1879 | 731957 | 889556 | 15945 | 48.80 | 40.15 | 9/11 | 27652 | 9.07 | 3442 | 1.13 | 6587 | 2.16 | 37681 | 12.31 |
| 1880 | 794906 | 941855 | 16119 | 45.27 | 38.34 | 9/10¼ | 27700 | 8.34 | 5631 | 1.69 | 6105 | 1.84 | 39436 | 11.87 |
| 1881 | 824148 | 970649 | 16617 | 45.13 | 38.35 | 10/2½ | 32666 | 9.50 | 5825 | 1.70 | 4451 | 1.29 | 42942 | 12.44 |
| 1882 | 838209 | 986135 | 16383 | 43.78 | 37.21 | 10/7¼ | 33909 | 9.71 | 6012 | 1.72 | 5361 | 1.54 | 45282 | 12.97 |
| 1883 | 899907 | 1103703 | 19016 | 47.33 | 38.59 | 11/- | 37178 | 9.92 | 5824 | 1.55 | 5584 | 1.49 | 48586 | 12.96 |
| 1884 | 910094 | 1143146 | 18386 | 45.25 | 36.03 | 11/1½ | 36951 | 9.75 | 6282 | 1.66 | 6579 | 1.73 | 49812 | 13.14 |
| 1885 | 931356 | 1190380 | 19220 | 46.23 | 36.17 | 10/10¾ | 38496 | 9.92 | 8827 | 2.28 | 8371 | 2.15 | 55694 | 14.35 |
| 1886 | 954191 | 1211030 | 20250 | 47.35 | 37.46 | 10/4¼ | 39452 | 9.92 | 9433 | 2.38 | 9439 | 2.37 | 58324 | 14.67 |
| 1887 | 960242 | 1262438 | 21979 | 49.86 | 39.00 | 9/10½ | 39683 | 9.92 | 7989 | 2.00 | 8416 | 2.10 | 56088 | 14.02 |
| 1888 | 994396 | 1299342 | 23603 | 51.76 | 40.69 | 9/10 | 41645 | 10.05 | 7086 | 1.71 | 7112 | 1.72 | 55843 | 13.48 |
| 1889 | 1007681 | 1307485 | 24383 | 52.78 | 41.77 | 10/7¼ | 45607 | 10.86 | 6649 | 1.59 | 6445 | 1.53 | 58701 | 13.98 |
| 1890 | 1074927 | 1390598 | 25826 | 52.38 | 41.60 | 12/8 | 47890 | 10.69 | 6495 | 1.45 | 7425 | 1.66 | 61810 | 13.80 |
| 1891 | 1207558 | 1559037 | 28629 | 51.72 | 41.13 | 14/2¾ | 56488 | 11.23 | 7342 | 1.46 | 7271 | 1.44 | 71101 | 14.13 |
| | | | | | | | | | | | | 1.36 | | |
| 1892 | 1188959 | 1549403 | 28628 | 52.51 | 41.39 | 13/11¾ | 54331 | 10.97 | 6707 | 1.36 | 7242 | 1.45 | 68280 | 13.78 |
| 1893 | 1136978 | 1469108 | 25789 | 49.40 | 39.32 | 12/3¾ | 50127 | 10.58 | 5908 | 1.25 | 6829 | 1.44 | 62864 | 13.27 |
| 1894 | 1187155 | 1575258 | 28275 | 51.93 | 40.21 | 11/3 | 49587 | 10.03 | 6566 | 1.33 | 7102 | 1.43 | 63255 | 12.79 |
| 1895 | 1183392 | 1553680 | 27737 | 51.10 | 39.99 | 11/6½ | 49394 | 10.02 | 6730 | 1.36 | 6940 | 1.41 | 63064 | 12.79 |
| 1896 | 1190489 | 1552800 | 26999 | 49.38 | 38.95 | 11/1¼ | 48857 | 9.85 | 6773 | 1.37 | 7166 | 1.44 | 62796 | 12.66 |
| 1897 | 1238441 | 1616959 | 27951 | 49.07 | 38.72 | 10/10½ | 50545 | 9.80 | 6929 | 1.34 | 7146 | 1.44 | 64890 | 12.58 |
| 1898 | 1257494 | 1651220 | 28195 | 48.75 | 38.25 | 11/1½ | 51387 | 9.81 | 7433 | 1.42 | 6984 | 1.33 | 65804 | 12.56 |
| 1899 | 1316195 | 1720292 | 29760 | 49.21 | 38.75 | 11/9½ | 54405 | 9.92 | 7476 | 1.36 | 7283 | 1.33 | 69164 | 12.61 |
| 1900 | 1318311 | 1763004 | 30424 | 50.22 | 38.66 | 13/5¾ | 57297 | 10.43 | 7191 | 1.31 | 7510 | 1.37 | 71998 | 13.11 |
| 1901 | 1322114 | 1749165 | 30610 | 50.37 | 39.20 | 15/10½ | 63573 | 11.54 | 7224 | 1.32 | 7742 | 1.40 | 78539 | 14.26 |
| 1902 | 1309261 | 1695351 | 30415 | 50.50 | 40.19 | 13/2½ | 59377 | 10.89 | 7331 | 1.34 | 7511 | 1.38 | 74219 | 13.61 |
| 1903 | 1311499 | 1724181 | 31202 | 51.48 | 40.54 | 12/8½ | 58814 | 10.76 | 7883 | 1.45 | 7245 | 1.32 | 73942 | 13.53 |
| 1904 | 1310818 | 1731085 | 31028 | 51.27 | 40.17 | 12/7¾ | 58645 | 10.74 | 8207 | 1.51 | 7146 | 1.30 | 73998 | 13.55 |
| 1905 | 1323915 | 1758823 | 32198 | 52.60 | 41.01 | 12/5½ | 58474 | 10.60 | 7411 | 1.34 | 7144 | 1.30 | 73029 | 13.24 |
| 1906 | 1334565 | 1774974 | 32882 | 53.26 | 41.50 | 11/9¾ | 58052 | 10.44 | 7526 | 1.35 | 6977 | 1.26 | 72555 | 13.05 |

**One aspect** of Whitaker's time as Locomotive Superintendent was his desire to collect data for evaluation. At the time, statistical analysis was in its infancy, but he, for one, realised its importance, perhaps having been coached in these techniques whilst at Derby. Surviving records give some hint of the depth of his scrutiny and are recorded here in this summary of mileage, coal consumed, and expenditure incurred year by year between 1876 and 1906. It is interesting to note that whilst mileages achieved more than doubled during this period, costs per mile dropped from 14.62p in 1876 to 13.05p in 1906 suggesting how effectively the railway was being run by comparison to the days of the S&D. (Author)

## ROLLING STOCK.

|  | 1875 | 1905 |
|---|---|---|
| **Engines** | | |
| Passenger Engines | 14 | 23 |
| Goods " | 11 | 53 |
| | | 76 |
| Duplicates | " | 8 |
| Total Engine Stock | 25 | 84 |
| **Carriages** | | |
| First Class (including Saloons) | 12 | 6 |
| Second " | 16 | " |
| Composites | 19 | 32 |
| Third Class ( " " ) | 15 | 89 |
| Horse Boxes | 12 | 14 |
| Carriage Trucks | 10 | 10 |
| Passenger Vans | 12 | 22 |
| Milk Vans | " | 18 |
| | 96 | 191 |
| Duplicates | " | 16 |
| Total Carriage Stock | 96 | 207 |
| **Wagons** | | |
| Cattle Wagons | 65 | 135 |
| Goods Covered Wagons | 40 | 190 |
| Goods and Coal Wagons | 583 | 820 |
| Timber Wagons | 60 | 72 |
| Ballast &c Wagons and Vans | 12 | 31 |
| | 760 | 1248 |
| Goods Vans | 18 | 40 |
| Travelling Crane and Wagons | 1 | 4 |
| Gas Holder Trucks | " | 2 |
| | 779 | 1294 |
| Duplicates | " | 36 |
| Total Wagon Stock | 779 | 1330 |

## STAFF RETURN.

| Station | 1875 | 1905 |
|---|---|---|
| Burnham | " | 4 |
| Highbridge | 187 | 320 |
| Bridgwater | " | 6 |
| Glastonbury | " | 1 |
| Wells | 4 | 6 |
| Bath | 51 | 162 |
| Radstock | 2 | 24 |
| Chilcompton | " | 1 |
| Shepton Mallet | 2 | 1 |
| Evercreech Junction | " | 2 |
| Cole | 1 | " |
| Templecombe | 12 | 28 |
| Sturminster | 1 | " |
| Blandford | " | 1 |
| Wimborne | 9 | 8 |
| Bournemouth | " | 16 |
| | 269 | 580 |
| Clerks | 3 | 11 |
| Superintendent | 1 | 1 |
| Total Staff | 273 | 592 |

**Two more** examples of the information regularly gathered by the Locomotive Superintendent's department – here in summary form from 1905. The loco fleet has trebled over thirty years, the number of carriages virtually doubled, whilst wagons have enjoyed a smaller increase. Meanwhile Whitaker's staff has more than doubled in size. (Author)

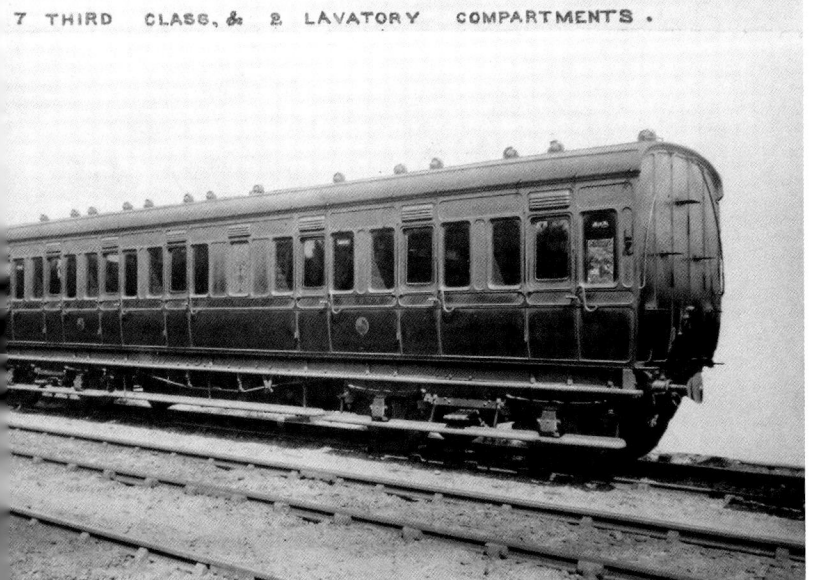
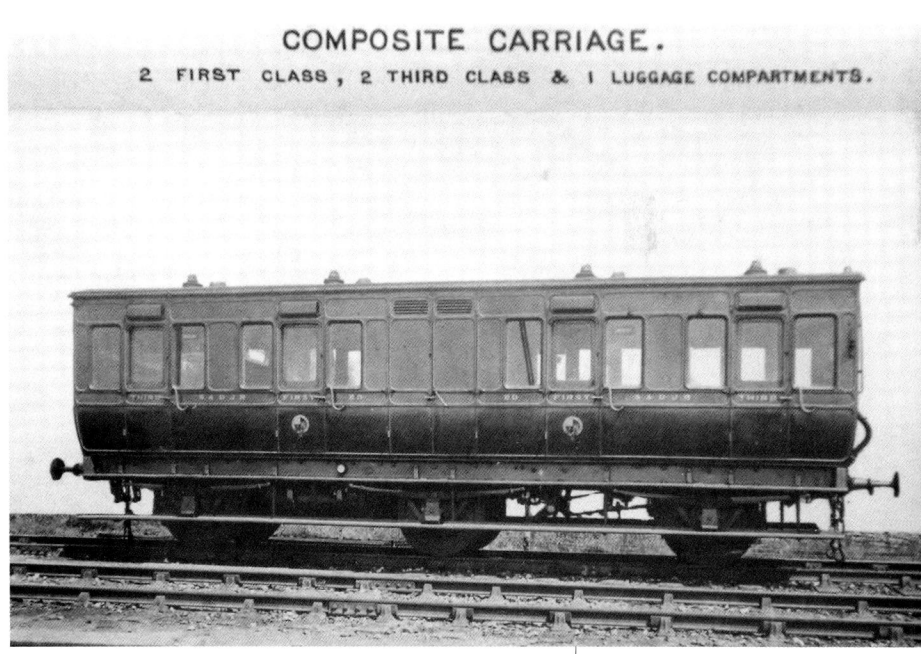

**Robin Atthill** noted that carriage construction was carried out by the S&D from its earliest days and they were painted chocolate brown. There was a gradual evolution in design, reflecting practice elsewhere in the industry, but due to the nature of this line, non-corridor types prevailed for some time. There were carriages with four then six wheels, then those with bogies which appeared in 1898. Local production appears to have ended in 1913, by which time Midland Railway stock began to proliferate. However, there was still a place for the old S&D carriages which were painted blue to match the colour of the engines by then. Examples of the different carriages produced by the company are featured above. (Author)

**Bath Station** in the early years of the twentieth century with a mixed bag of carriages on display. (Author)

**(Top photo)** By 1918 Midland Railway carriages, from through trains mostly, were a common sight on the line. By this stage, these comprised clerestory type coaches (middle left), in this case a wooden bodied steel framed corridor version designed by David Bain and his team (he was then the Carriage and Wagon Superintendent). During 1917, this was supplemented by a plain roofed version designed by Robert Whyte Reid (middle right), then the company's Works Manager of the Carriage Department. The bottom photo captures a typical scene at Bath with a 4-4-0 S&D engine pulling a rake of Midland carriages. This locomotive (number unclear) was a Samuel Johnson designed 5ft 9in Class engine rebuilt with an 'H' type round topped boiler between 1907 and 1911. (Author)

## A Line Created and Remembered • 37

HORSE BOX. STANDARD. 18'-0 WITH LOCKER.

CATTLE WAGON. 8 TONS.

HIGH SIDE WAGON. 8 TONS.

Although remembered primarily for his success as Locomotive Superintendent, Whitaker used his engineering skills in many other areas. One cause of the Radstock rail disaster was the lack of control over trains travelling over single track sections and this issue remained a constant thorn in the side of safe and efficient running. Under Dykes there had been some doubling of track, but this solution wasn't always practical and in some sections single running remained and with it a risk of collision. By 1905, these sections could be found between Bath Junction and Midford, Templecombe and Blandford and Broadstone and Corfe Mullen. Where single line running was necessary, a tablet system had been introduced to ensure that two trains couldn't operate over the same single section of track at the same time. As a result, an Edward Tyer designed and patented electrical tablet system, which first appeared in 1878, began to be installed, which contributed in no small way to the S&D's improved safety record.

**As one** would expect of a largely rural railway the transport of livestock and farm produce lay at its core and the wagons used, as the pictures produced above reveal, reflected this trade. (Author)

**However, the** Somerset coalfields were still heavily worked and created much needed revenue for the company. The picture above captures a typical load pulled by an 0-4-0 tank locomotive (No. 25A) at Writhlington Colliery, on the edge of Radstock, at the turn of the century. This colliery finally closed in 1973, surviving the railway by seven years. (D. Neal)

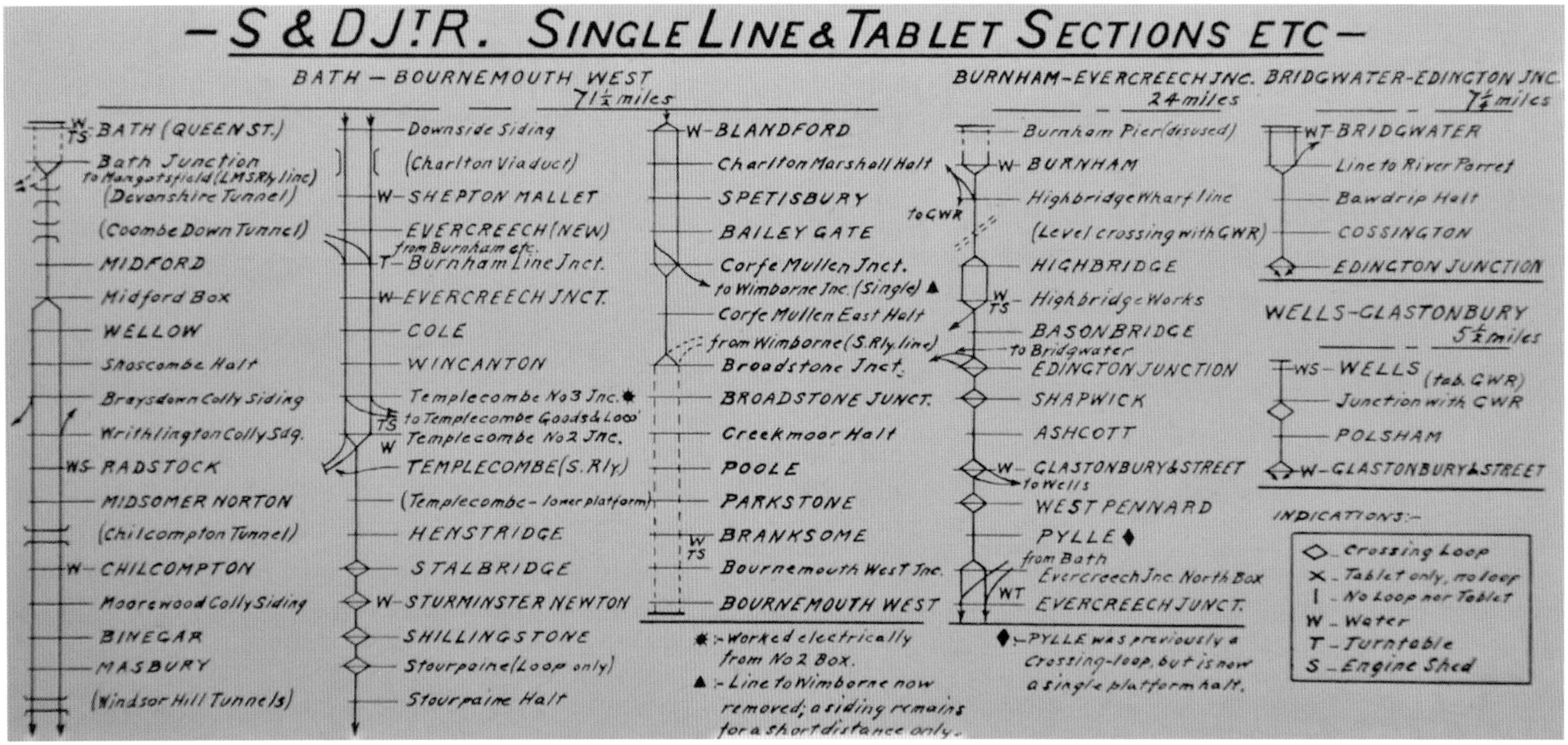

**A simple** schematic to show the extent of single lines on the S&D. Poor management of the system led to the collision at Radstock on 7 August 1876 and the need for tighter control. As a consequence, a programme of track doubling was undertaken and where it wasn't feasible, a tablet system was gradually installed.

While this was a significant development it had one disadvantage. Picking up a tablet meant that trains had to slow down to 10mph or stop completely to allow an engine's crew to gather up the leather pouch containing the tablet held out by a signalman. This method ensured there was a time penalty, particularly for express services, and automation, to some degree or other, was a desirable goal in speeding up the process. Whitaker put his mind to the problem, studied various solutions tried by other companies. However, he soon discovered that they still required a restriction on speed if they were to operate effectively, added to which, they operated with a fixed arm on locomotives positioned at right angles to the track and this had safety implications. So, he set about designing his own exchange apparatus which after use swivelled through 90 degrees until it was parallel to the track and couldn't foul lineside equipment or passengers. Trials over the Bridgwater Branch, and on experimental posts set up at Highbridge, proved successful and installation work commenced in 1905. In due course, nine stations, near single track sections, and seventy-one locomotives were equipped. A year later, the company were able to report that the Whitaker system was shaving appreciable amounts off journey times, without compromising safety.

**A crucial** but time consuming practice – slowing down to collect a tablet at Sturminster Newton before entering a section of single track in 1898. Here the crew of a Johnson designed 4-4-0 gather in a leather pouch containing a tablet with its large metal hoop to aid pick up. (D. Neal)

**One of** many photographs taken during the exchange mechanism trials at Highbridge. In due course Whitaker patented the design.
(D. Neal)

As an example, they reported that seven minutes had been taken off the time taken for the 2.10 pm Bath to Bournemouth luncheon express to travel non-stop between Green Park and Poole.

The early years of the twentieth century were peaceful enough for the S&D. Traffic volumes remained stable and the improvements wrought by Dykes and Whitaker brought many benefits, not least in the safety of its operation. But their time was drawing to a close. Dykes retired in 1901 and died seven years later at his home in Bath. In 1911, he was followed by Whitaker, having been succeeded by Mervyn Ryan, who enjoyed a long retirement, dying aged 92 on 5 March 1938, also in Bath. However, this peace was illusory. Events were moving fast in Europe, then the rest of the world, as long periods of conflict, interspersed with economic recession and social depression, became the norm. It was a time of massive change that would transform Britain beyond all recognition and with it the S&D.

The beginning of the Great War in 1914 was welcomed by many, but it would prove to be a short-lived celebration as casualties soon rose from hundreds of thousands into millions. The S&D, like all railways across Britain, soon found demand for its services increase. The need to carry extra goods was a clear necessity, but troop movements also became a regular task. However, the conflict brought with it a change of management. Increasingly aware of the need to co-ordinate the work of the railways and move beyond a natural rivalry between all the companies then operating, the government brought in centralised control under its Railway Executive Committee. This carried over to post-war years and led to a major review of the industry's structure, it being concluded that there were far too many companies for effective management. These deliberations ended with the 1921 Railways Act which gathered the majority of these companies, including the S&DJR, into four groups. The S&D would disappear and be absorbed jointly by the successors to the Midland and LSW railways – the London Midland Scottish and the Southern Railway. These changes came into force on 1 January 1923. Nevertheless, their impact was not felt immediately, and things went on pretty much as before, events largely controlled by the economic depression that followed the war and the hardships this caused.

Although the post-war recession was chiefly responsible for the trading difficulties that all railway

**For some** time, the need to provide stronger engines to tackle the S&D's steep gradients was of paramount importance. Thought focussed on the 0-8-0 configuration for a time, but then followed the example of George Churchward and the GWR and adopted the 2-8-0 layout instead. Churchward's Class 2800 first appeared in 1903 and other companies soon followed this example. The Midland Railway, with Henry Fowler as Chief Mechanical Engineer, supported by James Clayton, his Assistant Chief Locomotive Draughtsman, produced a 7F 2-8-0 that became synonymous with the S&D. Six were built at Derby in 1914 and another five would be added in 1925. The timing couldn't have been better and these six engines did sterling service throughout the Great War. Here engine No. 85, which appeared in August 1914, is undergoing main line trials near Elstree in March 1918. (D. Neal)

**In peacetime,** the influence of the 7Fs continued. From 1923, this would be in the guise of the LMS, as captured in this photo. (Author)

**In due** course, the LMS's more powerful and numerous 8F 2-8-0s would become a common sight on the S&D as this photo taken towards the end of the line's life on 3 March 1966 at Writhlington demonstrates. (Author)

companies faced, there was another factor to be considered. The Great War had encouraged the birth of mass movement along Britain's fledgling road system. Tens of thousands of military vehicles were sold as surplus post-1919. Although basic by later standards, they were sufficiently cheap and reliable to encourage those eager to build their own transport businesses to take the first step. However, at this stage, the threat they posed to the railway's dominance was barely registering, but this would soon change and competition quickly grow and with it the quality of roads over which they could pass. For the moment, though, any threat posed to passenger traffic was a small one and would remain so until employment and incomes rose sufficiently to allow people to purchase their own cars in large numbers. Either way, the writing on the wall was becoming clearer and the bigger railway companies would soon struggle to make a profit and survive let alone the smaller businesses such as the S&D. And so the story of the next forty years would be one of decline, fruitless quests to increase trade, efforts to make their operation as cheap as possible and, when all failed, closure. Throughout the remaining years, the S&D's staff remained loyal and attempted to make the railway pay, but in the long run, all this endeavour was doomed to failure.

In any business, simple economics cannot be ignored for long. For the S&D, the decline each year was measured by the company's accounts, with costs exceeding revenue by varying amounts. Savings measures

**When this** photograph was taken at Wellow Station in 1920, of engine No. 54, an 1885 Vulcan Foundry built 0-4-4T, passenger traffic had returned to normal after the disruption of war (this locomotive was withdrawn from service in November that year). Regular passenger traffic from the north to the south coast across the S&D had also been restored, in particular regular services from Manchester to Bournemouth, which had begun in 1910. From the beginning, it appears to have been an unnamed service, but in 1927 it acquired a name that became synonymous with the S&D – the Pines Express – which ran over the Mendips until re-routed in September 1962. (D. Neal)

were adopted and, in 1930, this led to some centralisation of management functions within the LMS and the Southern Railway. One significant loss was the Traffic Superintendent's post and the team that sat around him in Bath. Some of the staff were re-employed locally, others made redundant, while some re-located to Derby or London. With their going, the sense of autonomy, which the railway had enjoyed since its inception, finally appeared lost. In truth, it had probably been an illusory independence once the leasing agreement with the Midland Railway and LSWR had taken effect.

At the same time that this was happening, the Southern Railway took over responsibility for all engineering matters. With these changes, the need for a local Locomotive Superintendent and Resident Engineer was considered and, in due course, both posts also disappeared. In the place of these three terminated senior managers, the LMS's District Controller, S. Sealy, became, to all intents and purposes, the S&D's manager. In the months that followed, he was given the title Assistant Divisional Superintendent. It was a post that appears to have survived, in various guises, until 1958, by which time all semblance of independent management across Britain's railways had gone with nationalisation and the creation, in 1948, of British Railways. But in the meantime, the cuts went on.

The works at Highbridge were all but closed in 1930, with the loss of 300 or so jobs, but some did find employment with other companies. Three years later, the

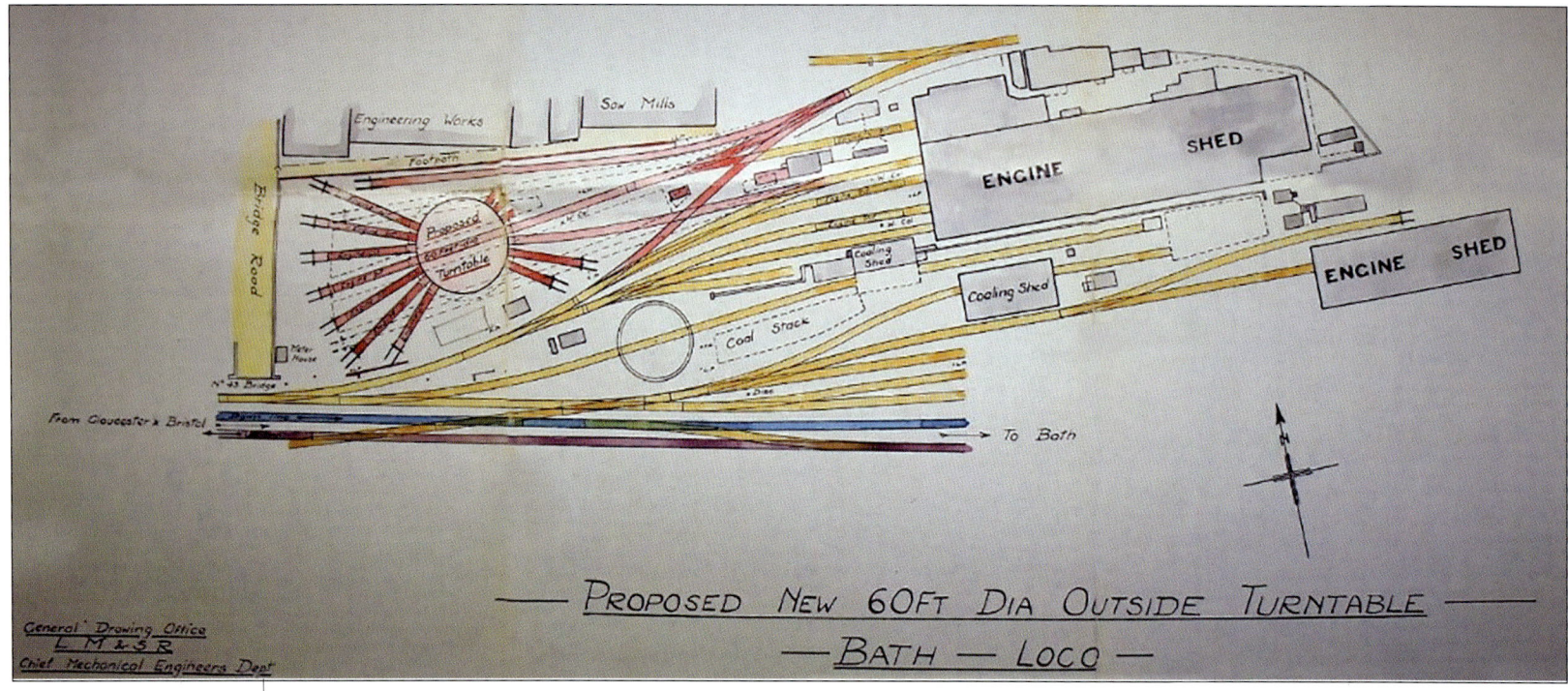

PROPOSED NEW 60FT DIA OUTSIDE TURNTABLE — BATH — LOCO

**The turntable** at Bath, although adequate for the size of locomotive employed when first installed, was too short to accommodate the 7F 2-8-0s in 1914. To get round this problem, tender first running proved necessary on many services. In the 1930s, a proposal to change it for a 60 foot version reached fruition, as witnessed by this drawing. Approval followed and replacement went ahead. (Author)

final elements of the Somerset and Dorset's shipping fleet were paid off, though the wharf at Highbridge remained active and was used by ships from other companies. All this was accompanied by other cuts and reductions in service, not least the closure of the Wimborne line. And yet these savings didn't improve the railway's fortunes to any extent and, as some commentators believed, only postponed the inevitable. By then, the country had drifted into another world war and for six years, economic pressures were put to one side. Instead, the country faced invasion, the ravages of a bombing campaign and a U Boat onslaught that sought to strangle Britain's will to resist. Whether the line made money or not was largely immaterial, now it was a military supplier in a region crucial to future campaigns as the Allied powers fought back and then prepared for D Day and the advance across France, Belgium and Holland into Germany's heartland.

For the most part, the line wasn't directly affected by enemy action but did become collateral damage on at least two occasions. As part of the Baedeker raids, Bath was targeted on 25-27 April 1942, and sustained much damage in the process. In all, 417 people were killed and another 1,000 injured. The Green Park area received a numbers of hits and the station buildings were slightly damaged, though not as severely as many other sites around the city. Five months later, a bomber dropped its load on Templecombe Station in an attempt to disrupt traffic on the Southern's main line to the west and over the S&D's metals. Any damage was quickly repaired.

Working parties across Britain were by then very experienced in these matters, and both lines remained open. Just as well, because they were fulfilling essential roles. As Robin Atthill recorded, 'traffic was heavy throughout the war years, with the emphasis on freight, day and night…millions of tons of equipment and raw material being lifted over the Mendips by the famous Class 7s'.

Then it was all over and peace, when it came, offered few prospects for the railways except their gathering together as a government-led body. Post-1948, the line was assigned to the Railway Executive's Southern Region, with day to day supervision vested in a District Traffic Superintendent based in Southampton. Under him sat an Assistant District Operating Superintendent located at Bath,

**The changes** introduced by the Transport Act of 6 August 1947 saw BR take control of the S&D. Over the next two decades, this resulted in a plethora of different locomotives operating over the line. These included ex-LMS, ex-Southern and later on, when the Western Region absorbed the line, ex-GWR engines. To these were added the product of BR's standardisation programme. For the lineside enthusiast, this made for a fascinating day out. These two photos capture the variety of engines that could regularly be seen. (Left) A Southern Region Pacific in company with a BR Standard Class 2 tank engine in the process of backing down from Templecombe on to the S&D line to the west of the station. In the above picture two ex-LMS 8Fs sit at rest in a shed at Bath. (Author)

responsible for all motive power. However, he was a London, Midland Region man, and so a strong link was maintained with the old Midland Railway and LMS. This carried on until 1950, when boundaries were redrawn and the Western Region gained a commercial interest in the line northwards from Cole, with the Southern still managing all operations. As a result, the LM then transferred some locomotives to the Southern, which also took on direct responsibility for the railway's Motive Power Depots at Bath, Templecombe, Branksome, Radstock and Highbridge. With that, the Midland's eighty-year involvement with the S&D appears to have ended.

Change is an ever-present fact of life, none more so than in the life of BR at this time. After years of financial stringency caused by recession and the accumulative effects of wartime overuse and reduced maintenance, the railways remained in a parlous state. Road transport during the 1950s was also providing significant opposition. The fictional railway portrayed so well in the film *The Titfield Thunderbolt* captured the essence of the problems which the S&D in particular faced. In the film, a group of villagers fight back with some success against the modernising world, just as their forbears had tried to stop the spread of the railways in the early nineteenth century. The reality then could not

have been more different. Now the railways, still for the most part relying on steam, did not have the weapons to fight back. But still attempts to find an answer went on. Nevertheless, without the wherewithal to fund a massive modernisation programme, or for that matter the political will to do so, the car and the lorry increased their domination of Britain's transport system.

As this happened, BR continued to tinker around the edges of the problem. In 1958, regional boundaries were again re-drawn, with the Western Region assuming control of the line from Bath to Henstridge, managing this new acquisition from Bristol. Then, in January 1963, the Western

**The 'enemy'**, in the form of cars and lorries, is at the gate as the S&D's life finally comes to an end in 1966. Around the still impressive Bath (now named Green Park) Station, all is changing, with new buildings going up and soon this famous landmark will be a simple adjunct to a Sainsbury's store. (Author)

took control of the old LSWR line between Salisbury and Exeter and with it a new boundary with the Southern was drawn. Nearly ninety years after the GWR had sought to control the S&D by leasehold, the descendants of that company finally had their way. However, growth or survival were no longer goals, instead the line became a 'withered arm of the Western Region – a moribund and hopelessly uneconomic branch line', as Robin Atthill described it. Once in place, the Western's managers seem to have followed a deliberate policy of starving the line of business. And, in due course, Richard Beeching, by then BR's Chairman, and his team reviewed the need for the Bath to South Coast link, found little justification to do so and the services that remained were gradually withdrawn. In 1964, a Labour government came to power with a promise to curtail more cuts, soon confirming the S&D's closure, despite impassioned appeals for it to remain open. And in March 1966, the end came.

It is difficult to compress the history of the S&D into only a few pages because there are so many aspects of its life to describe – the people who made it run, the communities it served, its operation, the types and variety of locomotives and rolling stock that graced its metals, the social and economic history of the times and much more. And yet its life can be summed up quite briefly. The product of an ambitious plan by men of substance during the 1850s and 1860s, though probably based on the flawed assumption that such a rural, sparsely populated area could support such a project. By dint of

**The S&D's** blue painted locomotives and carriages are a distant memory, but a BR standard locomotive pulling Southern Region green carriages still conjures up a wonderful image of a 'delightful' line just before its life came to an end. (Author)

constant expansion to create a line from Bath to Bournemouth, it was hoped that financial success would follow. With excessive debt to be managed, this proved impossible and only by leasing the line to two large, established companies was collapse avoided. Until the Great War, life went on, the railway survived, but then recession followed the conflict and the decline began. Different solutions were tried, but it was a downward spiral and eventually the line could not survive the apparent antipathy of British Railways Western Region, who assumed control in 1958. Were they right to allow it to happen? Accountants would undoubtedly have said yes at the time, but the world and opinions change and now a different view might be taken. However, there is no going back. Too quickly, and perhaps too hastily, the line and its infrastructure were destroyed by BR's Western Region, in an act, some believe, of petty tyranny. And, with this, the S&D passed into history.

## Chapter 2
# LOCOMOTIVES IN THE EARLY YEARS (1854 to 1875)

By any standards, the locomotives that found their way on to the S&D during its lifetime were a mixed bag, reflecting the many companies that had some influence over its day to day operation. Sometimes this was simply a matter of convenience, the company seeing its adjoining neighbours – the Bristol and Exeter Railway and LSWR – as a ready source of locos and rolling stock to populate the line as it slowly grew. Then, later on, the engines available reflected the wishes of its lease holders and, of course, there were times when the S&D's directors felt able to invest in new engines themselves, adding a new element to their fleet. As time passed, the LMS, having absorbed the Midland Railway in 1923, brought their influence to bear on motive power, as did BR post-1948, when locomotives trickled in from the other regions in a random sort of way, or so it seemed. Yet at no time in its history was the S&D's fleet, although an incredibly varied one, ever large.

The S&D never, it seems, sought to design or build its own locomotives as other companies did. However, there was, when the occasion arose, some limited planning ability which enabled specifications to be prepared when expressions of interest were being sought through tendering processes. So, in the early days of both the Somerset Central and Dorset Central, there was a reliance on neighbouring companies to provide the engines and rolling stock they needed if they were to operate at all. In due course, specialist manufacturers, such as George England and Co, Vulcan and Fox, Walker and Co, were commissioned to build the locomotives for them, but these were only in small numbers. Then, of course, there was the question of gauge to consider. The Somerset section was constructed using Brunel's 7ft 0¼in gauge track, which made sense because of its links with the Bristol and Exeter Railway at Highbridge. And, through a leasing agreement, the B&ER agreed to provide locomotives and rolling stock until 1861, effectively tying the SCR's hands for the foreseeable future in the matter of motive power, so putting off the day when they would have to procure their own.

Meanwhile, in Dorset, where the line interconnected with the LSWR, 4ft 8½in gauge track had been adopted to allow a crucial link to be established. When the northern and southern lines were eventually linked, a third rail was added to the broader gauge sections, allowing through working. While this was happening, the question of a single, universally accepted gauge was being addressed by parliament. In due course, this long running, and contentious standardisation debate was finally resolved in favour of narrower gauge track. It would take until 1892 for the last vestiges of broader gauge to disappear, but the Somerset and Dorset achieved changeover much earlier in 1870, so could focus solely on standard gauge engines from then on.

So the S&D was rarely master of its own destiny when it came to locomotives and relied heavily on other companies to design and build the engines it needed for its entire life. However, this policy did allow them to pick and choose to a certain extent and even experiment at times, which isn't a bad route to take when finances and the chance of added investment are limited. As a result, the evolution of its locomotive fleet is the story of the design efforts of other men and other companies, until British Railways came into existence and a new standardisation programme became the order of the day. And the names this policy threw up, in varying degrees, reads like a who's who of Britain's railway history – James Pearson, John Gooch, Joseph Beattie, William Beattie, William Adams, Samuel Johnson, Richard Deeley, Henry Fowler, William Stanier, Tom Coleman, Oliver Bulleid, Robert Riddles and Ron Jarvis amongst others. So the S&D, no matter how modest in scale and ambition, could boast some remarkable antecedents, as well as a host of designers employed by the many private manufacturers that existed at the time on whose work they drew periodically.

When the Somerset Central line opened in 1854, there was only a small requirement for locomotives and rolling stock from the Bristol and Exeter Railway. Six trains a day between Highbridge and Glastonbury was hardly excessive and unlikely to be difficult to service. The B&ER had Brunel as resident engineer and the first section was open to traffic in 1841. From then until 1849, it was reliant on the GWR for its locomotives, but with James Pearson as its

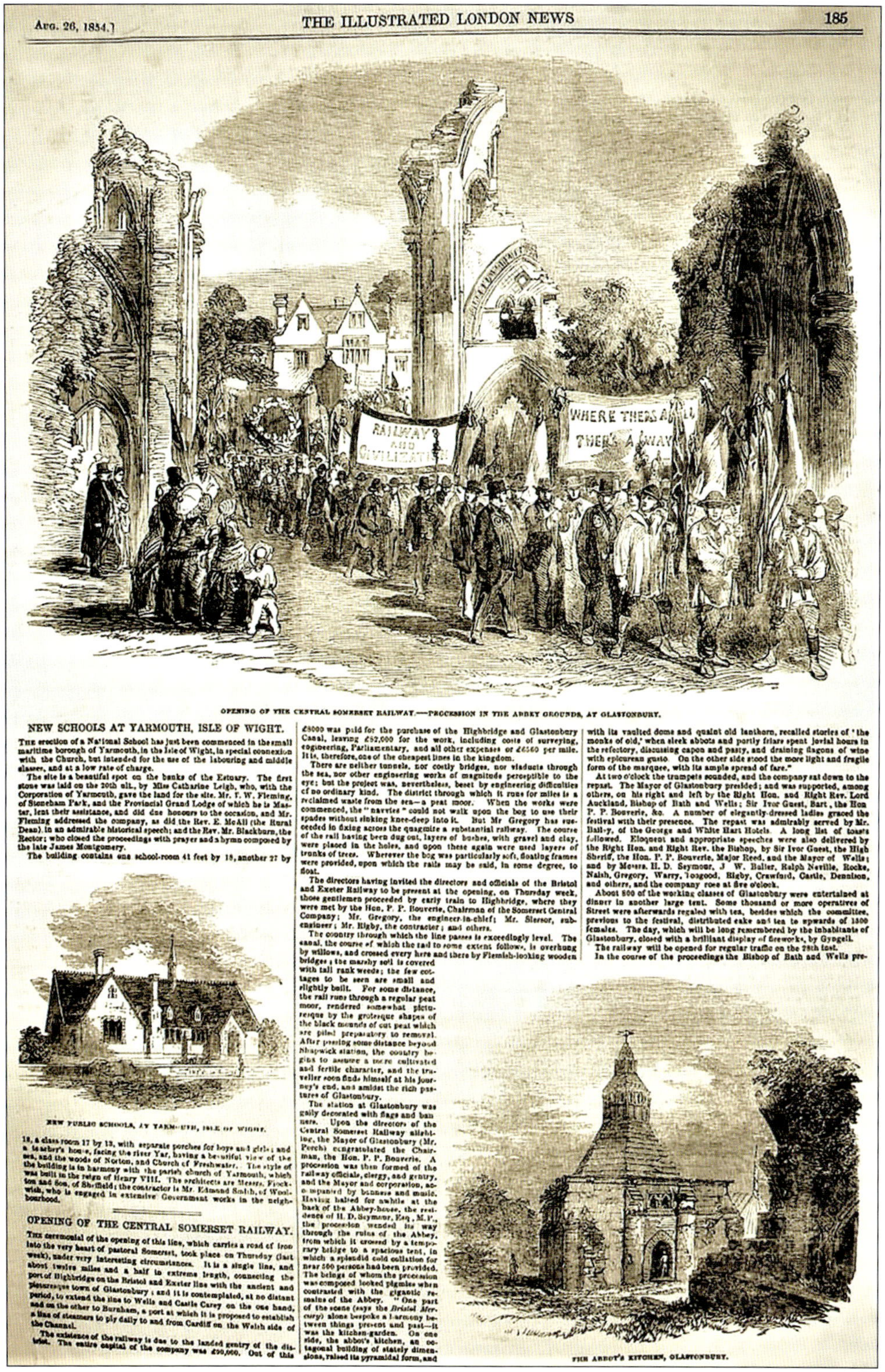
Aug. 26, 1854.] THE ILLUSTRATED LONDON NEWS 185



OPENING OF THE CENTRAL SOMERSET RAILWAY.—PROCESSION IN THE ABBEY GROUNDS, AT GLASTONBURY.

NEW SCHOOLS AT YARMOUTH, ISLE OF WIGHT.

The erection of a National School has just been commenced in the small maritime borough of Yarmouth, in the Isle of Wight, in special connexion with the Church, but intended for the use of the labouring and middle classes, and at a low rate of charge.

The site is a beautiful spot on the banks of the Estuary. The first stone was laid on the 20th ult., by Miss Catharine Leigh, who, with the Corporation of Yarmouth, gave the land for the site. Mr. T. W. Fleming, of Stoneham Park, and the Provincial Grand Lodge of which he is Master, lent their assistance, and did due honours to the occasion, and Mr. Fleming addressed the company, as did the Rev. E. McAll (the Rural Dean), in an admirable historical speech; and the Rev. Mr. Blackburn, the Rector; who closed the proceedings with prayer and a hymn composed by the late James Montgomery.

The building contains one school-room 41 feet by 18, another 27 by

NEW PUBLIC SCHOOLS, AT YARMOUTH, ISLE OF WIGHT.

18, a class room 17 by 13, with separate porches for boys and girls; and a teacher's house, facing the river Yar, having a beautiful view of the sea, and the woods of Norton, and Church of Freshwater. The style of the building is in harmony with the parish church of Yarmouth, which was built in the reign of Henry VIII. The architects are Messrs. Flockton and Son, of Sheffield; the contractor is Mr. Edmund Smith, of Woolwich, who is engaged in extensive Government works in the neighbourhood.

OPENING OF THE CENTRAL SOMERSET RAILWAY.

The ceremonial of the opening of this line, which carries a road of iron into the very heart of pastoral Somerset, took place on Thursday (last week), under very interesting circumstances. It is a single line, and about twelve miles and a half in extreme length, connecting the port of Highbridge on the Bristol and Exeter line with the ancient and picturesque town of Glastonbury; and it is contemplated, at no distant period, to extend the line to Wells and Castle Carey on the one hand, and on the other to Burnham, a port at which it is proposed to establish a line of steamers to ply daily to and from Cardiff on the Welsh side of the Channel.

The existence of the railway is due to the landed gentry of the district. The entire capital of the company was £90,000. Out of this £6000 was paid for the purchase of the Highbridge and Glastonbury Canal, leaving £82,000 for the work, including costs of surveying, engineering, Parliamentary, and all other expenses or £6560 per mile. It is, therefore, one of the cheapest lines in the kingdom.

There are neither tunnels, nor costly bridges, nor viaducts through the sea, nor other engineering works of magnitude perceptible to the eye; but the project was, nevertheless, beset by engineering difficulties of no ordinary kind. The district through which it runs for miles is a reclaimed waste from the sea—a peat moor. When the works were commenced, the "navvies" could not walk upon the bog to use their spades without sinking knee-deep into it. But Mr Gregory has succeeded in fixing across the quagmire a substantial railway. The course of the rail having been dug out, layers of bushes, with gravel and clay, were placed in the holes, and upon these again were used layers of trunks of trees. Wherever the bog was particularly soft, floating frames were provided, upon which the rails may be said, in some degree, to float.

The directors having invited the directors and officials of the Bristol and Exeter Railway to be present at the opening, on Thursday week, those gentlemen proceeded by early train to Highbridge, where they were met by the Hon. P. P. Bouverie, Chairman of the Somerset Central Company; Mr. Gregory, the engineer-in-chief; Mr. Slessor, sub-engineer; Mr. Rigby, the contracter; and others.

The country through which the line passes is exceedingly level. The canal, the course of which the rail to some extent follows, is overhung by willows, and crossed every here and there by Flemish-looking wooden bridges; the marshy soil is covered with tall rank weeds; the few cottages to be seen are small and slightly built. For some distance, the rail runs through a regular peat moor, rendered somewhat picturesque by the grotesque shapes of the black mounds of cut peat which are piled preparatory to removal. After passing some distance beyond Shapwick station, the country begins to assume a more cultivated and fertile character, and the traveller soon finds himself at his journey's end, and amidst the rich pastures of Glastonbury.

The station at Glastonbury was gaily decorated with flags and banners. Upon the directors of the Central Somerset Railway alighting, the Mayor of Glastonbury (Mr. Porch) congratulated the Chairman, the Hon. P. P. Bouverie. A procession was then formed of the railway officials, clergy, and gentry, and the Mayor and corporation, accompanied by banners and music. Having halted for awhile at the back of the Abbey-house, the residence of H. D. Seymour, Esq., M.P., the procession wended its way through the ruins of the Abbey, from which it crossed by a temporary bridge to a spacious tent, in which a splendid cold collation for near 500 persons had been provided. The beings of whom the procession was composed looked pigmies when contrasted with the gigantic remains of the Abbey. "One part of the scene (says the *Bristol Mercury*) alone bespoke a harmony between things present and past—it was the kitchen-garden. On one side, the abbot's kitchen, an octagonal building of stately dimensions, raised its pyramidal form, and with its vaulted dome and quaint old lanthorn, recalled stories of 'the monks of old,' when sleek abbots and portly friars spent jovial hours in the refectory, discussing capon and pasty, and draining flagons of wine with epicurean gusto. On the other side stood the more light and fragile form of the marquee, with its ample spread of fare."

At two o'clock the trumpets sounded, and the company sat down to the repast. The Mayor of Glastonbury presided; and was supported, among others, on his right and left by the Right Hon. and Right Rev. Lord Auckland, Bishop of Bath and Wells; Sir Ivor Guest, Bart., the Hon. P. P. Bouverie, &c. A number of elegantly-dressed ladies graced the festival with their presence. The repast was admirably served by Mr. Bailey, of the George and White Hart Hotels. A long list of toasts followed. Eloquent and appropriate speeches were also delivered by the Right Hon. and Right Rev. the Bishop, by Sir Ivor Guest, the High Sheriff, the Hon. P. P. Bouverie, Major Reed, and the Mayor of Wells; and by Messrs. H. D. Seymour, J. W. Bailer, Ralph Neville, Rocke, Naish, Gregory, Warry, Toogood, Rigby, Crawford, Castle, Dennison, and others, and the company rose at five o'clock.

About 800 of the working classes of Glastonbury were entertained at dinner in another large tent. Some thousand or more operatives of Street were afterwards regaled with tea, besides which the committee, previous to the festival, distributed cake and tea to upwards of 1500 females. The day, which will be long remembered by the inhabitants of Glastonbury, closed with a brilliant display of fireworks, by Gyngell.

The railway will be opened for regular traffic on the 28th inst.

In the course of the proceedings the Bishop of Bath and Wells pre-

THE ABBOT'S KITCHEN, GLASTONBURY.

**This well-known** lithograph and article were published by *The Illustrated London News* in 1854 to commemorate the opening of the Somerset Central Railway. It carefully contrasted the old World, with the ruins of the Glastonbury's Abbey as a backdrop, and the new with the words on the banners and the optimism they convey. But the SCR, built on a shoestring, wasn't in a position to buy or build its own locomotives. So, for many years, they relied on the broad gauge Bristol and Exeter Railway Company and any locomotives or rolling stock they could spare. It was a cobbled together solution that reflected the way the railway was run until it closed in 1966 with a myriad number and type of engines gracing the line. (D. Neal)

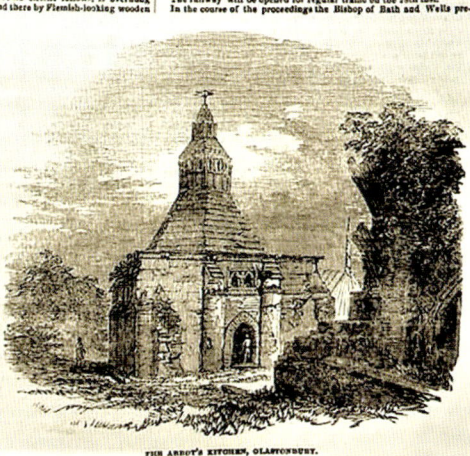

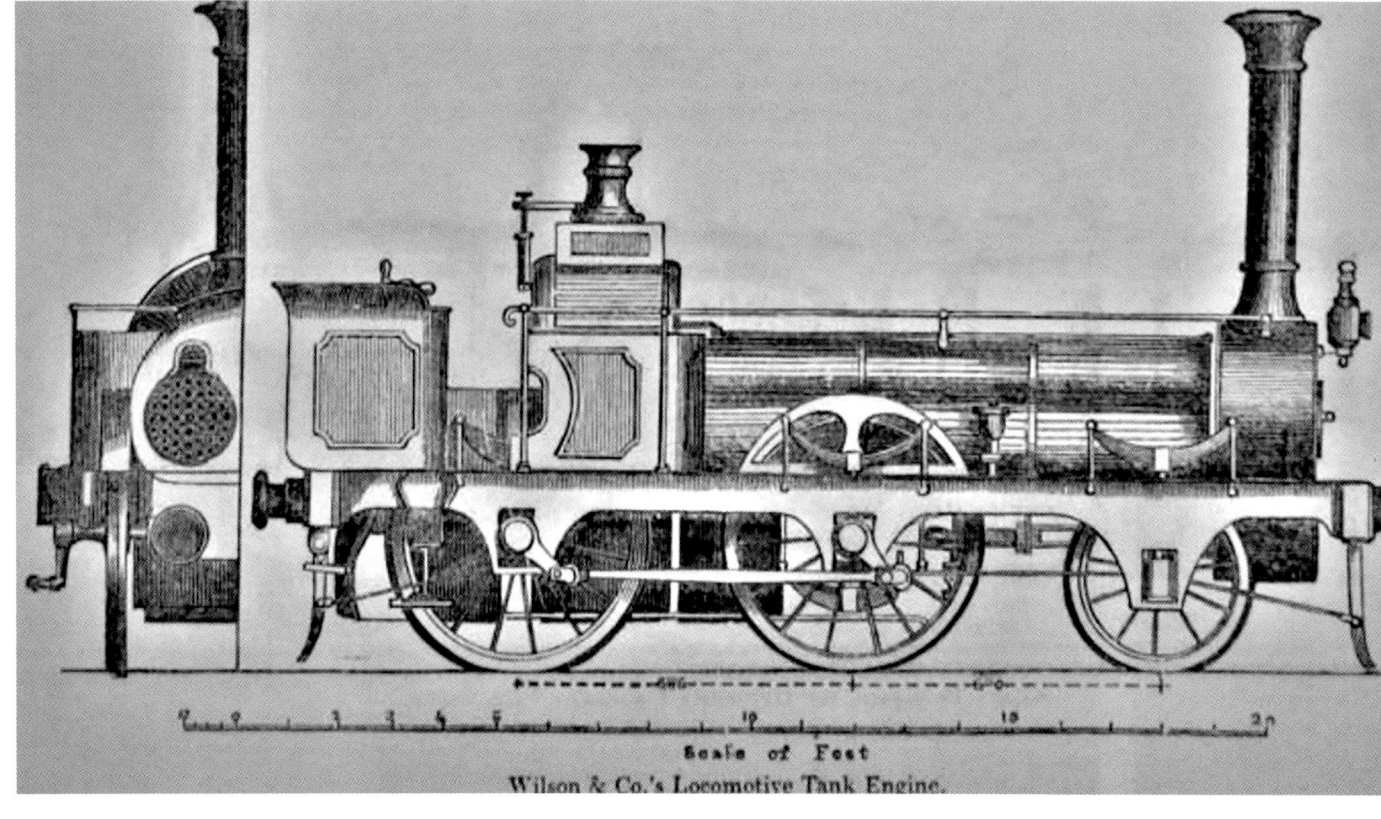

**A typical** product of E.B. Wilson and Co, suppliers to many companies including the Bristol and Exeter in the mid-nineteenth century. This is one of the few drawings that appears to have survived giving a general impression of the type of engine the company produced and which would have been a familiar sight on the B&ER's branch lines and, possibly, the Somerset Central. (Author)

Locomotive Superintendent they began designing and building their own engines. So it would have been a mixture of GWR and B&ER engines that would have serviced the newly opened Somerset Central.

Pearson was born on 29 March 1820 to James and Alice in Blackburn. Having successfully passed through school, he qualified as a civil engineer. From here, he found employment in the rapidly expanding railway industry. In 1850, he was appointed to the B&ER, a post initially based in Exeter but then moved to the company's new works on the Bath Road site in Bristol which opened in September 1854. He appears to have been a creative man who directly involved himself in the design of locomotives and soon left his mark on the railway's construction as well as its procurement programmes. However, it wasn't until 1859 that the first of the thirty-five locomotives for which he can claim credit appeared. By this stage, the company were beginning to hedge their bets when it came to gauge and only twenty-three of these engines were of the 7ft 0¼in variety, while ten were the narrower gauge. But before this programme commenced, the B&ER acquired a number of tank locomotives built by E.B. Wilson and Co of Leeds and Rothwell and Co of Bolton for use across its system, but more specifically on its branch lines. Although few records of specific engines traversing the Somerset Central exist for this period, it is recorded that the first train over the line was pulled by a Wilson 2-2-2 No.33, one of three owned by the B&ER. Then on 17 May 1857, a Rothwell 4-4-0 saddle tank locomotive No. 47, one of six built in 1855, was derailed in the sidings at Glastonbury. It is probably safe to assume that these were only two of the locomotives that worked the line at this time, with other types being used as required. Over the years there has been some speculation over what these might have been, but the choices are limited by the fairly small number of tank engines then available to the B&ER. Here, the work of Pearson, in conjunction with Wilsons and Rothwells, is important.

When it came to designing new engines, Pearson produced a range

**A Rothwell** and Co built locomotive for the Bristol and Exeter. These 2-2-2 class engines appear to have been simple but sturdy designs, ideal for branch line use. (Author)

BRISTOL AND EXETER ENGINE OF 1859

of tank engines more substantial in size than those that had gone before. In particular, there were his 4-2-4T engines built with three different sized flangeless driving wheels. There were eight locomotives with 9ft diameter wheels, built by Rothwell's to a Pearson design in 1853/54, then two with 7ft 6in wheels and four with 8ft 10in wheels built by the B&ER in the years 1862 to 1868. All of these engines had two supporting bogies, inside plate frames, brakes to all wheels, well and back water tanks and domeless boilers. In practice, the engines with 9ft wheels could, when descending long inclines, reach speeds marginally in excess of 80mph. However, for reasons that aren't entirely clear, the 8ft 10in locomotives were seen as replacements for four of the '9 footers'. It was a small reduction in wheel size, but one Pearson thought necessary, suggesting they may have held the track better, gave the engines wider route availability or simply improved their riding quality. It is impossible to say with any certainty whether these engines appeared on the Somerset Central, alongside or in place of the 2-2-2s and 4-4-0s already mentioned, whilst the broad gauge track remained in place. However, it remains an interesting possibility.

The Dorset Central Railway was in a similar position to their Somerset neighbours and relied upon the LSWR for engines and

*Right*: **One of** the B&ER's Pearson designed, Rothwell built broad gauge 4-2-4Ts (No.44) which appeared in 1853/54 as the Somerset Central was opened. It is difficult to confirm whether these substantial tank engines found their way on to the branch line, but it remains a possibility due to the limited number of engines that the B&ER had available, suitable for such duties. Either way, they would have been a very familiar sight at Highbridge during the early years of the S&D. It is rumoured, but not confirmed, that it is James Pearson standing in front of this locomotive. (D. Neal)

*Left*: **During 1868,** three Pearson 4-2-4Ts appeared with 8ft 10in driving wheels, a fourth being added in 1873, here represented by engine No. 2002. They were built to replace an equal number of engines with 9ft wheels, which were withdrawn from service in 1868 and 1873 (two on each occasion). (Author)

*Opposite below*: **Although LSWR** engine No. 57, *Meteor*, isn't recorded as having operated over the Dorset Central's line but three of her sisters did – No. 53 *Mazeppa*, No. 58 *Sultan* and No. 61 *Snake*. All ten engines of this class were built at Nine Elms in 1847 with some remaining in service until 1872. It is recorded that they operated mostly over the Templecombe to Cole section presumably until the end of their lives. (D. Neal)

## Locomotives in the Early Years • 53

rolling stock to run services over its 4ft 8½in metals between Wimborne to Blandford from 1 November 1861. By luck, we find that some of the engines that worked the line have been recorded. The first of these was a single member of the ten-strong Bison Class 0-6-0 No. 49 itself called *Bison*. These engines were built between 1845 and 1848 at Nine Elms and would join three engines of the Mazeppa 2-2-2 Class – *Mazeppa* (No. 53), *Sultan* (No.58) and *Snake* (No. 61) – that tended to operate the Templecombe to Cole section of the line. These 2-2-2s were built, with their seven sisters, at the 'Elms' in 1847 under the direction of John Viret Gooch, the LSWR's Locomotive Superintendent until 1850. Then came engines of three classes built following Joseph Hamilton Beattie's succession to the Superintendent's post. In 1851, within months of his arrival, the first of fifteen Hercules Class 2-4-0 engines rolled out of Nine Elms, with production running on until 1854. In due course, one of these, No. 41, *Ajax*, would regularly work over the Dorset Central line. This engine would be joined by

**(Left) John** Viret Gooch (1812-1900), the LSWR's Locomotive Superintendent until 1850. He was one of three brothers who contributed much to locomotive design – Daniel with the GWR and Thomas with the Manchester and Leeds Railway. On leaving the LSWR, he joined the Eastern Counties Railway. His designs featured in the early years of the Dorset Central Railway. (Right) Equally important was the work of Joseph Hamilton Beattie (1808-71), the Irish born engineer who served as the Locomotive Superintendent until dying from diptheria in 1871. By this time, he had designed an extensive range of 2-4-0 and 0-6-0 engines, with his 0298 Class 2-4-0 well tanks probably being the best remembered today, with two passing into preservation. He was succeeded as Superintendent by his son William. It was recorded that Beattie was 'one of the most versatile and exuberantly inventive men that ever headed a locomotive department. He turned out gadgets and dodges both weird, wonderful in a bewildering spate. He ruled everything relating to mechanical engineering on the LSWR by inflexible decree, and having invented things, he proceeded to build them'. (D. Neal)

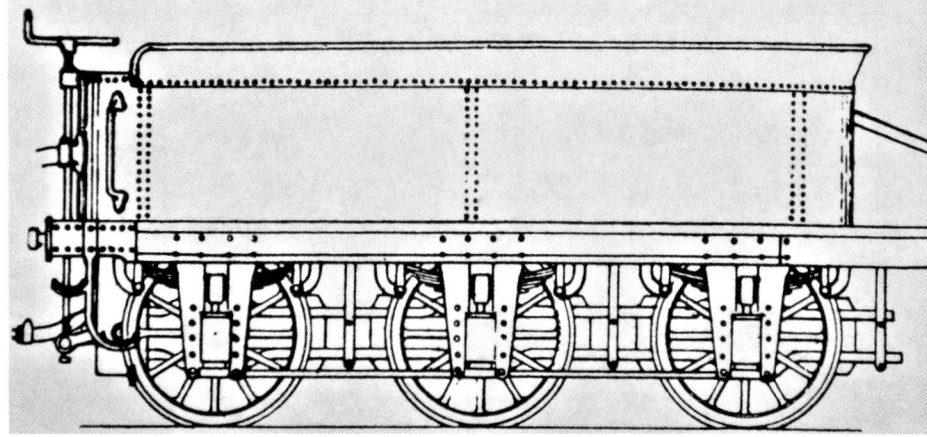

**Hercules Class** 2-4-0 tender engine No. 44 as built at Nine Elms between 1851 and 1854. This locomotive's sister, No. 41 *Ajax*, was recorded as operating on the Dorset Central's line between Wimborne and Blandford. All members of the class, which had 5ft 6in driving wheels, were gradually withdrawn from service between 1875 and 1884. (Author)

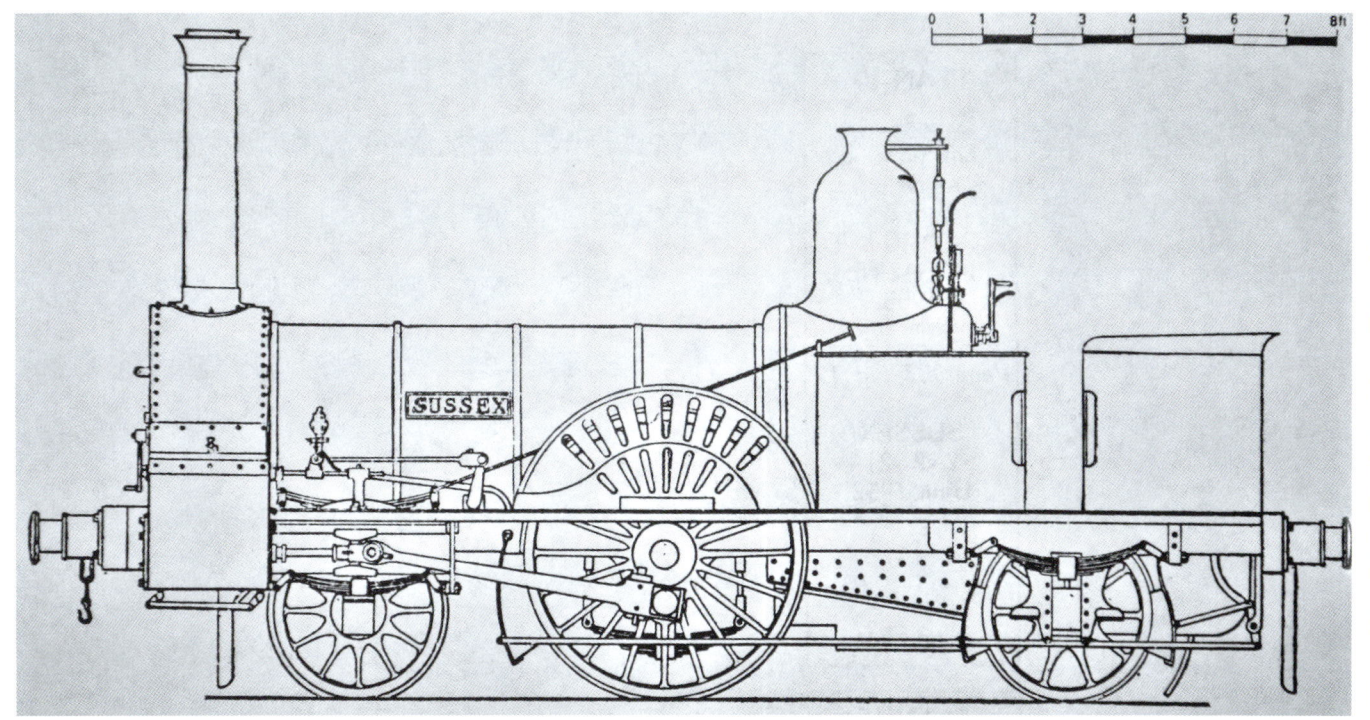

**Joseph Hamilton** Beattie produced eight Sussex Class 2-2-2 well tank engines in 1852 at Nine Elms, one of which, No. 15 *Mars*, made regular appearances on the Dorset Central's line. (Author)

**The first** member of the Sussex class is captured in this photograph taken in 1870 outside the old passenger station at Nine Elms. Two years later, the engine was scrapped. The reverse of this print records that Joseph Beattie is in the group – possibly the figure in the middle with the top hat and long coat. If so, he would soon depart the scene, dying in 1871. (Author)

**Only three** members of Beattie's Nelson Class of 2-4-0 well tank engines were built in 1858. It fell to engine No.145, *Hood*, to represent the type on the Dorset Central. The last of class survived until 1885. (D. Neal)

**The Bison** Class of 0-6-0 tender engines, here represented in diagrammatic form by No. 106 *Panther* and photographically by No. 51 *Elephant*, were ten in number. By the time No. 49, *Bison*, appeared on the Dorset Central line, they had been in service for more than fifteen years and were beginning a rebuilding programme that would see them become members of the Lion Class. The photo of *Elephant* shows this engine in modified form, a state she reached in 1863 at the same time as *Bison* was rebuilt at Nine Elms. *Panther* became a Lion in December 1865. In modified form these engines continued in service until 1887. It is possible that No. 49 saw service on the Dorset Central in both guises. (D. Neal)

single members of two other classes. The first of these was No. 15 *Mars*, one of eight 2-2-2 Sussex Class engines built in London in 1852. This was followed by No. 145, *Hood*, one of three Nelson Class 2-4-0 well tank locomotives built in 1858 again at Nine Elms.

The amalgamation of the Somerset Central and Dorset Central on 1 September 1862 proved to be a watershed in many ways, not least in the future of the locomotive fleet. Although the two lines were not yet joined, this was only a matter of time. Work on the Templecombe to Blandford section was underway and would finally connect both railways during 1863. In addition, and with the aid of the LSWR, access to Poole was achieved allowing the coast to coast link to become a reality and, hopefully, boost trade. The question of different gauges remained, but this does not seem to have presented insuperable problems to the directors. However, for a time it meant that there were two locomotive programmes running side by side and would stay this way until the gauge issue was resolved. In anticipation of this the SCR had, by 1862, begun a programme of adding a third rail to its system to allow dual running.

For the standard gauge section of the Dorset line the LSWR would continue to provide locomotives, many of which were the latest products of Pearson and Nine Elms. It was a policy that seemed to work, but meant that the LSWR took 60 per cent of the gross receipts as agreed in its five-year deal with DCR. But the Somerset Central, with dual running over their section of track becoming a reality, and before amalgamation took place, initiated a procurement programme to provide the standard gauge engines and rolling stock it now needed. It was a decision largely driven by the end of its leasing agreement with the B&ER during 1861 and the conversion of the track to dual running. In the background to these deliberations was the statutory requirements contained in the 1848 Gauge of Railways Act. This sought to make the 4ft 8½in gauge the universally accepted standard across Britain, but it was a solution that presented the GWR with huge difficulties having invested so heavily in broad gauge across its growing network.

It was a business decision, but also one of principle, the company's managers genuinely believing in the superiority of this gauge over the standard version. And so the GWR fought a tenacious rear guard action that was probably doomed the moment the 1848 Act passed into law, but it was a fight they found difficult to abandon. They would eventually lose and be faced with the massive costs of conversion, but whilst they fought, they were unlikely to sympathise with anyone seeking to distance themselves from their ideas. In due course, this resulted in the GWR's takeover of the Bristol and Exeter, when it seemed they might break ranks. A similar fate could have befallen the SCR, then the Somerset and Dorset when formed in 1862. However, they managed to remain independent of the GWR and the Bristol and Exeter and began investing in their own locomotives and rolling stock.

The directors and the Resident Engineer, Charles Gregory, moved quickly to produce drawings and specifications for locomotives and rolling stock. Having limited experience of design and operation, he enlisted the help of the B&ER's James Pearson in developing these plans. In due course, tenders were sought from leading manufacturers of the day. Rothwells, Beyer Peacock of Manchester, Slaughter Gruning of Bristol and George England and Co responded, but, surprisingly, two well established companies did not reply. Why Kitsons of Leeds and Robert Stephenson & Co chose not to participate in the bidding process is unclear. It may simply have been that the order was too small, or they were already fully loaded with other work. In assessing the bids, the SCR went for the lowest price and awarded a contract for eight 2-4-0 engines at £1,850 each, plus £1,800 for six tenders to George England, with delivery set for July 1861. In March, the order was amended. It was decided that one of the tenderless engines should be fitted with a water tank, which added £150 to the bill.

George England and Co, like most industrial concerns at the time, started in a small way during 1839. The proprietor began his career as an apprentice to John Penn, a marine engineering firm in Deptford on the south bank of the Thames. Aged 28, he set himself up in business and rented a small factory in Hatcham and here he experimented, then developed his own designs, some being turned into patents. Of particular note were a weaving machine and a traversing screw jack, but in due course he turned his attention towards the design and construction of locomotives.

**George England** (1811-78) as he appeared in the 1860s when his Hatcham Iron Works was in the process of producing locomotives for the Somerset Central Railway. Contemporary accounts suggest that he was a tough, no nonsense autocrat, even by the standards of the age. This probably motivated his staff to work hard but it was, by all accounts, a harsh regime which eventually caused his work force to strike in 1865. It is said that this so damaged the company's reputation that it never fully recovered. Nevertheless, the locomotives the works produced seem to have found favour in the industry. (D. Neal)

**Photographs of** S&D engines when the line was in its infancy are few and far between and sadly, as in this case, in a very battered and bruised state. In 1861, George England and Co of London were commissioned to build eight 2-4-0 locomotives for the company. The first seven, of which the engine pictured below is the fourth, were all delivered by November. They would retain their SCR markings for a very short period only, becoming S&DR locomotives in September 1862. Engines 1, 3, 4 and 5 remained with the S&D until November 1874/75 when part exchanged with Fox, Walker and Co of St George's Bristol, as part of a deal to acquire five 0-6-0 small industrial tank engines. In due course, Nos. 1 and 5 went to the Bishop's Castle Railway, while it is thought that Nos.3 and 4 may have gone to the East & West Junction Railway. It seems that 2, 6 and 7 remained with the S&D until withdrawn and scrapped, No. 6 having been converted into a side tank engine in 1876, being given the number 26 in the process. (Author)

In 1849, he appears to have sold his first engine and two years later his locomotive named *Little England* won a gold medal at the 1851 Great Exhibition in London. This helped considerably in underpinning his growing status in the industry and by 1861, when awarded the SCR's contract, his business had expanded exponentially and with it the size of his Hatcham Iron Works.

By July 1861, work had progressed fairly well and the first four engines and six tenders were ready for delivery. Unfortunately, the SCR were unable to accept them and they ended up in sidings at Salisbury under the temporary care of the LSWR, where they were joined by the remaining four engines. They could have languished there much longer but for Robert Andrews, the newly appointed Locomotive Superintendent, who issued instructions to have them test run to make sure they were fit for purpose. This work began in October, allowing one locomotive to be transferred to Templecombe for use as a works engine assisting in the construction of the line to Cole. A month later, the other locomotives were also moved from Salisbury into the SCR's safe keeping where they worked from Templecombe to the Somerset Coast, whilst No. 8 saw service on the Burnham branch line. Very soon, this engine was found to have insufficient water capacity for the purpose for which it was intended. As a result, the decision was taken to fit a 705 gallon saddle tank and the engine returned to traffic in March or April 1862 in this form. Meanwhile, a seventh tender was ordered from Hatcham Iron Works. Why it didn't feature as part of the initial order is unclear. However, it may simply have been

**The second** of eight engines George England and Co built for the SCR, now with S&DR markings following amalgamation in 1862. The early form of bent plate weather board over the footplate is interesting and shows a growing concern for the welfare of the crew, albeit a very small one. In working order, these engines weighed 25 tons 4 cwt and were equipped with 15in x 18in diameter cylinders, with leading wheels of 3ft 6in and 5ft coupled wheels. The wheelbase was 13ft 8in, plus tender. The boiler was flush-topped and 9ft long, with a diameter of 3ft 10in, and was attached to a firebox measuring 4ft 10in in length with a grate area of 13sq ft (here there is some uncertainty though. Atthill records the grate area as being 14½sq ft, whilst in 1938, *Locomotive Magazine* settled on 13sq ft). Together, they produced a total heating surface of 781sq ft and a working pressure of 115lb. Another noteworthy feature is the locomotive's framing – outside when it came to the coupled wheels, inside for the leading wheels, which, as can be seen, were sprung below the running plate. The four wheeled tenders had outside framing and are thought to have been constructed in a similar way to GWR tenders of this period. The locomotives were not equipped with brakes, with the tender having a handbrake which applied wooden blocks to the front and rear of both the left hand side wheels. (D. Neal)

a savings measure. When in service, at least one locomotive would be undergoing maintenance at any one time so would not require a tender. A juggling act perhaps that larger companies, with many more locomotives of any one class in the servicing cycle, might successfully accomplish. But on a much smaller railway harder to achieve and so a one for one ratio would have proved necessary.

The delivery of these locomotives presaged by only a year or so amalgamation which became a physical reality when the Blandford to Templecombe section was completed on 31 August 1863 and opened to regular traffic a month later. A single motive power policy was now an issue which the new unified board had to address. The Dorset Central, as we have seen, relied on the LSWR for locomotives and rolling stock, while the SCR had begun to develop its own procurement plan. This was now extended but would remain at the mercy of the new company's parlous financial state through its fourteen years of existence.

In a spirit of optimism, the newly created S&D ordered two more 2-4-0 tender engines from George England in December 1862 at a cost of £2,550, which, it seems, included both engine and tender. These locomotives, numbered 9 and 10, were broadly the same as the other engines built at Hatcham, but they did include some modifications that took account of experience gained with the other locos. The two cylinders were increased in size to 16in x 18in and the firebox, which was of the raised top variety, had a grate area rather larger than the earlier engines. Together, these changes were probably sought to help increase power output on the line's steeper gradients. Nevertheless, they could only produce 7,066lbs of tractive effort compared to that of 8,602lb generated by engines 1 to 8. In addition to this, the total wheelbase was increased by 4 inches, although the coupled wheelbase was larger by 7 inches. There were also some modifications to the frames and cabs, in an effort to give the footplate crew slightly more protection. In this modified state, both locomotives entered service in August 1863.

These two engines were quickly followed by a single tank locomotive, which had distinguished itself by being built for exhibition purposes. According to some accounts, it appeared at the 1862 International Exhibition in South Kensington, alongside George England's traversing screw-jack. However, information to confirm this is limited, the events catalogue simply listing an engine with tender being displayed, which isn't to say that they were joined as one. However, in the months after the exhibition, it was offered for sale to a number of potential customers without success and was finally snapped up by the S&D for £1,800 in October 1863. Here it remained, as No. 11, until sold in 1870, but to whom remains unclear. One account suggests the Admiralty for use at Sheerness, another that it became the property of the LSWR and eventually ended up working on the Lee-on-Solent Light Railway.

The next four locomotives purchased by the S&D were again 2-4-0 tender engines. With trade over the line gradually improving, now that the coast to coast link was complete, the company felt sufficiently confident to invest in their locomotive fleet. An order was placed on Hatcham in November 1863 for these engines at a cost of £2,600 each. However, England failed to meet the June 1864 delivery date and the engines finally arrived in September.

Although similar to engines 9 and 10 there were a number of differences. While the overall wheelbase remained the same at 14ft, the distance between the coupled axles was increased to 8ft 2in, although this was compensated for by a reduction to 5ft 2in between the leading and first driving axle. This was partially the result of a revised firebox arrangement. It remained the same length, but its width was increased, and its raised top continued round the sides to connect with the frames. They also carried D.H. Clarke's patented smoke consuming apparatus as an extra not listed in the original order or specification. When a royalty payment became due in 1865, it was forwarded to England's, who were then sued by the patent holder for this and other unpaid fees. In the melee that followed, the S&D's directors, perhaps sensing an opportunity to stagger the £10,400 payment for these engines, negotiated an initial cash amount of £1,200, then quarterly sums of £1,000. By this sleight of hand, and bearing in mind their own developing financial problems, payment was nicely delayed in their favour.

By juggling debt and payments in this way, the company's stretched finances were managed to a certain extent, though it seemed beyond the

**Engines 9** and 10 were built by George England & Co in 1862. Of note is the slightly more substantial cover provided for the crew, now fitted with side extensions. All these early 2-4-0s were comparatively small engines, even by the standards of the age. In each picture, the crew seem to be too large for their mounts making movement in the confined space of such limited footplates even more difficult. (D. Neal)

**The Machinery** Hall at the 1862 International Exhibition in which engine No. 11 was apparently displayed. Though offered for sale by George England and Co post-display to various customers, it struggled to find a buyer until the S&D made an offer of £1,800 the following year. However, its time with the company was short lived and it was sold in 1870. (Author)

**The single** 2-4-0 tank engine, No. 11, with a number of interested spectators gathered around, including footplate crew, station staff and possibly, on the far left, Robert Read, who played a leading role in the SCR, S&D's and S&DJR's history from 1853 to 1891. His tasks included Secretary, General Manager and, at one time, Managing Director. When built, it was painted blue and the engine garnered the rather fitting nickname 'Bluebottle'. However, the locomotive was soon repainted green, which in due course was applied to locomotives and rolling stock of the post-1876 Joint Railway. The locomotive appears to have been operated on the Glastonbury to Wells branch line for the duration of its stay on the S&D. (Author)

**A nicely** posed picture of engine No.13 with its extended canopy providing a little more protection for the crew. On 11 January 1866, this locomotive was involved in a collision on a section of single track near Wimborne Junction. A period of heavy snow and freezing conditions caused severe disruption to rail services across the West Country. However, the S&D managed to keep a skeleton service going, despite damage to signals, telegraph lines, compounded by frozen points. At 17.28 a much-delayed Burnham to Poole train, pulled by engines 2 and 15, arrived at Blandford. There a two-carriage train pulled by No. 13 was waiting to leave for Wimborne. To minimise the amount of traffic on the line it was decided to join both trains together, but the points couldn't be opened due to impacted snow to allow this to happen, so the Poole train departed by itself. Sadly, the guard on board failed to inform the stationmaster at Wimborne that engine No.13 was following and he let a Templecombe bound goods train, pulled by engine No. 9, proceed northwards. Meanwhile, at Blandford, the stationmaster there who should have waited to let the goods train pass failed in his duty and let No. 13 press ahead. In the snow and darkness, the two trains collided at a combined speed of about 25mph. Luckily, no one was killed and the injuries were fairly minor, but both locomotives suffered severe impact damage requiring extensive repairs at Highbridge. The Board of Trade recorded that 'the two engines were a good deal damaged as the cylinders, buffer beams and buffers were all broken, the inside frames bent, etc. The tenders were also a good deal damaged'. (Author)

directors to find a way to achieve a sounder base. Nevertheless, there was sufficient going on to justify buying more locomotives. The first of these was another 2-4-0 type, but this time an engine that started its life in 1842 as a 2-2-0T built by Bury, Edward and Co of Liverpool, shortly before they became Bury, Curtis and Kennedy. It was ordered and delivered to the South Eastern Railway. During 1857, it was rebuilt as a 2-4-0T for reasons that aren't clear and four years later it was bought by Joseph Firbank. At this stage, he was a railway contractor and he was prepared to pay £1,200 for an engine he could use to support construction work for the London, Brighton and South Coast Railway. Probably being surplus to his requirements, it was sold in October 1865 to George Reed, one of the S&D's directors and also a mine owner, for £1,450. Reed may have seen it as a useful addition to his colliery business, but instead sold it to the S&D with payment delayed for twelve months. It then seems to have found employment for a time in the Radstock area. When ownership was transferred the sales documents listed it as being 'fitted with a new copper firebox, a second hand set of brass tubes, new tyres, 14 in. cylinders and a spare six wheeled tender'. In reality, it proved too light weight for most duties and was eventually side lined at Highbridge Works and sold in 1876 for a peppercorn fee of £65.

This purchase was soon followed by two more engines from George England in November 1865. They were both 2-4-0s and were given the numbers 17 and 18, but they were significantly different from the types already acquired from

Hatcham. However, rather like the purchase of engine Nos. 11 and 16, this transaction owed more to opportunism than a sound business strategy. With a continuing shortage of locomotives, the company decided in 1864 to buy additional mixed traffic 2-4-0 engines, but due to lack of funds, the S&D wasn't in a position to place an order until August the following year with the Vulcan Foundry in Lancashire. And then its requirement for six new locomotives, at a cost of £2,900 each, couldn't be met by the company until mid-1866. Faced with a shortfall in motive power, the board decided to buy two engines 'off the peg', so to speak, and the S&D placed a 'wanted' advert in the railway press hoping to attract business that way. By chance, George England had been in the process of building twenty 118 Class 2-4-0 tender engines for the South Eastern Railway, but when strike action at Hatcham delayed the first deliveries the SER pulled out of the deal, fearing more postponements. However, whilst the solicitors wrangled, construction went ahead and fourteen locomotives were completed. In due course, as part of the negotiation, the railway agreed to take four of these engines at a cost of £2,400 each, but no others. Luckily, they managed to sell four of the others to the West Flanders Railway and two more to a company in Genoa. At this point, the S&D's need for extra engines was made public and they agreed to pay £7,000 for the last two locomotives.

The 118s first entered service in 1859 and during a long production run, 110 would be built by 1875. Sixty-eight were constructed at the SER's Ashford Works and the remainder by other manufacturers. With their unique two-sided fireboxes, they were something of an anomaly, but remained in service until 1897.

These engines were longer by 14 inches and heavier by more than five tons than the other 2-4-0s bought from George England. Their cylinders measured 16in by 24in., while the leading wheels were 4ft 6in and the coupled drivers 6ft. Their boilers were only marginally larger, but the new type of firebox was nearly four feet longer, with a grate area of 26sq ft by comparison to 14½sq ft in engines 1 to 8, which gave them a total heating surface of 979sq ft and a working pressure of 120lb. The frames were double in nature, with the outer pair being strengthened with separate hornplates and detachable tie bars. However, in service the frames proved to have a number of weak points and suffered some cracking. This increased the level of maintenance and repair required. Within a year or two, Highbridge had undertaken a strengthening exercise to reduce the amount of flexing and cracking taking place. It isn't known whether this corrected the problem completely, but it was often the case that frame defects were inherent to the design with many contributory causes. This wasn't their only shortcoming, though. Problems were experienced with the new firebox. Water

**In 1865** the S&D acquired two 2-4-0 engines built by George England to a James Cudworth design (118 Class) for the South Eastern Railway, of which Cudworth was Locomotive Superintendent, later becoming Locomotive Engineer. The S&D engines were numbered 17 (Right) and 18.
(D. Neal)

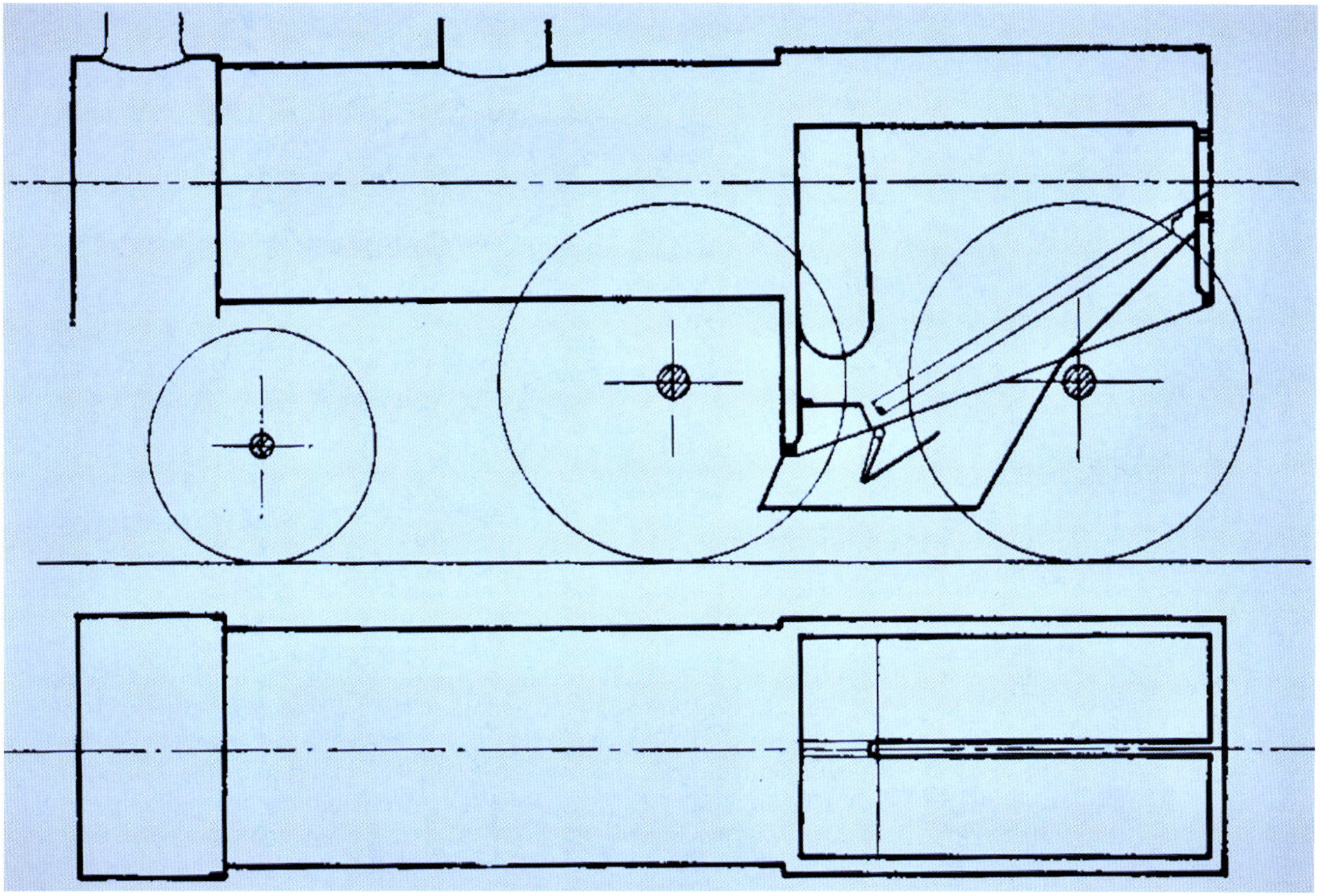

**A feature** of Cudworth's career was his continuing fascination with the reduction of smoke emissions from his locomotives. Burning coke rather than coal could have achieved this, but this alternative fuel was more expensive so made no sense economically. So, in 1859, he patented a new firebox design that burnt coal but reduced the smoke produced. As shown in this diagram, it contained a central, longitudinal partition containing water. As a result, the firebox had two separate firedoors, one on each side of the partition, and required a different firing technique – one side at a time. The theory was that as the coal burnt down on one side and it began to cool, the other side would be at peak temperature. As coal was then thrown into the cooler section the smoke produced would be burnt off by the greater heat generated in the other side of the firebox. It was a turn and turn about system that was found to work effectively, though was slightly more expensive to build and maintain than more conventional systems. It didn't find widespread use, though, and engines 17 and 18 were the only S&D locomotives to employ the Cudworth firebox. (D. Neal)

**In 1879,** Nos. 17 and 18 were rebuilt and appeared in this form. During 1895, No. 17 became No. 45. (Author)

**One of** the six Vulcan Iron Works built 2-4-0s for the S&D in 1866. With the company in the hands of the Receivers, the purchase of all these engines proved financially impossible. A compromise deal was agreed. The S&D accepted liability for the two already supplied and Vulcans agreed to seek buyers for the other four, which they did but it was a process that took another five years. The engines supplied to the S&D were given the numbers 19 and 20. This photograph is interesting because it shows No. 23 in S&D livery although never delivered to the railway. One wonders if they remained in this condition until sold to the German government in the hope that the S&D would find the cash to complete purchase. (D. Neal)

tightness was hard to achieve and leaks were a continuing problem, while the firebox seams often failed due to the heat. The six-wheeled tenders also suffered from weaknesses in their framing and had to be modified.

Although more powerful than their sisters on the S&D, any benefits gained from their greater pulling power were offset by higher operating and maintenance costs. These issues were ignored for a time, but as the engines grew older, the frequency of visits to workshops and higher running costs meant rebuilding or disposal. The S&D chose the former and by 1879 they were both rebuilt under the guidance of Midland Railway engineers. At this point, the boiler was replaced and the Cudworth fireboxes were removed and replaced with a brick arch version.

In October 1874, No.18 briefly made the news when involved in a fatal accident north of Evercreech Junction. It had just returned to traffic after a period of maintenance, during which the engine received new cylinders, tyres and crank axle, amongst other things. A few weeks later, it was heading to Bath on the newly opened extension, with a brake van and three carriages, when the wheels left the track just before Pecking Mill viaduct. The embankment had suffered a land slip following heavy rain and the track had sunk creating a serious dip which caused the engine to de-rail. Still with considerable momentum it crunched along the ballast for 15 yards or more and then crashed into the viaduct's right-hand parapet, which gave way. Momentum then carried it down the slope followed by the brake van, which had become detached from the carriages which stayed upright on the track above. Driver Carter was crushed to death, while the fireman, an inspector and the guard were injured, but luckily the passengers were unhurt although undoubtedly shocked by what had happened. Poor workmanship by the contractor who built the line and, in particular, lack of adequate drainage, coupled with a poor inspection regime were cited in the Board of Trade report as the main contributory factors. Damage to the engine was extensive but repairable and it returned to traffic a short while later.

Whilst the arrival of the two Cudworth designed engines gave the S&D a much-needed boost to its fleet, the need to acquire the six new locomotives from the Vulcan Foundry remained a pressing issue. But it was a programme quickly soured by the railway's stretched finances and their inability to raise capital through shares or debentures. By 1865, they had entered a period where they couldn't service their debts adequately and collapse was a likely outcome. The dichotomy they faced was only too clear. There was trade to be cultivated in the area, but it couldn't be procured because of lack of locomotive and rolling stock capacity. This undermined their ability to grow and for some months the company desperately sought a solution to their problems with little success. In June 1866, the collapse came and the Court of Chancery took charge of the S&D's affairs, appointed Receivers and quickly implemented a ruthless campaign of savings measures.

Meanwhile, the six new locomotives had, it appears, been completed by Vulcan, but in the prevailing climate the chances of the S&D being able to finance the purchase were slim indeed. With no cash available, the company sought to find other ways of acquiring the locomotives. Two had in fact been delivered on 14 July, but with restrictions on expenditure in place there was no immediate prospect of payment being made. Leasing these two engines for three years at a £1,000 per annum was suggested by the S&D, coupled with the suggestion that the other four be offered by the manufacturers for sale on the open market. Loath to make a loss or end up under capitalised themselves, Vulcan demanded full payment for the two supplied. However, they did agree to sell the others, provided that the S&D accepted liability for any loss up to £150 per engine. In 1871, they were finally sold to the German government for £10,000 leaving the S&D to pay £450 of the company's £1,600 loss on the deal.

Engines 19 and 20 were similar in many ways to George England supplied engines built to the S & D's specifications. The wheels

**The two** Vulcan built engines, now numbered 15A and 16A, as they appeared after their 1902/03 rebuilding when they received Midland Railway 'A' type boilers. When new, their outside frames were found to be weak and too easily damaged by the normal stresses of operation. (Author)

**For a** time towards the end of its life, 16A was immobilised at Highbridge and simply supplied steam to parts of the works there. (Author)

were the same size, with the same spring system over the leading wheels, the cabs were virtually the same and the fireboxes were raised. However, there were variations in the dimensions and output. The cylinders were larger at 17in x 22in, the engines were 5 inches longer, the boiler's diameter was 4ft 3in compared to 3ft 10in and longer by 8½ inches, the total heating surface was 1,097sq ft, 237sq ft greater than the earlier engines and they were 8 tons 6cwt heavier at 33 tons 10cwt. The tenders were of the six-wheeled variety and they could carry 1,575 gallons of water and more coal than the older England engines, which probably increased their range between top ups. In the course of their lives, they were rebuilt twice, firstly in 1880/81 then in 1902/03 under the auspices of the Midland Railway. Both survived until 1914, numbered 15A and 16A, after forty-eight years of service.

The arrival of these two engines marked the beginning of a moratorium on the acquisition of new motive power, despite the views of the company's managers and directors. With the Court of Chancery managing their business, there would have been little room for manoeuvre and virtually no flexibility. In the short term, such draconian measures might work, but eventually wear and tear will take effect making the engines and rolling stock less efficient no matter how good the standard of maintenance. But this wasn't likely to sway the Receivers, who were doing their best to keep the business going in any form and protect, as best they could, the interest of investors. It was a delicate balancing act which the directors sought not to upset, and which seemed to work. During 1870, the company was discharged by the court and regained control. Perhaps in the certain knowledge of this upturn, the running department quickly commissioned a report into the state of the locomotive and rolling stock fleet. The aim was to make plain the extent of any shortcomings and the impact this had on the ability of the company to operate. Frederick Slessor, the Resident Engineer, presented the findings to the Board on 31 December that year.

It made sorry reading. Of nineteen locomotives, one, No. 11, had just been sold, No. 8 had been 'laid aside' for conversion into a tender engine and No. 16 was in a static state and supplying steam at Highbridge Works. He then described the extent of downtime generated by repair and maintenance needs, adding a note about engine No. 17 with its 'unsatisfactory firebox'. His most telling comment, though, focussed on the inability of engines 1 to 10 to undertake main line duties due to their small size. He then went on to recommend that these underpowered locomotives be sold and replaced by 'six powerful six-coupled goods engines with large tenders'. By this stage, the long-harboured ambition to build the Bath extension was back on the table, a fact which Slessor didn't let pass. He added a footnote. Should the new line be built, the company would need eight larger passenger locomotives and four or more strong tank engines for banking duties. It is said that this report ruffled a few feathers amongst Board members, but Slessor's points were well made and didn't over or under estimate the scale of the problem. He also recommended that a Locomotive Superintendent, a post abandoned when the Receiver took control, be appointed as soon as possible. Since that date, Slessor had been covering this work and clearly felt that it now needed a full time occupant. Nevertheless, it took three more years before Benjamin Fisher was recruited from the Taff Vale Railway to fill the post. So the board clearly didn't feel able to acquiesce to Slessor's idea too quickly.

With the lifting of restrictions on the company, the directors were able to raise £160,000 in debentures immediately. Then, a year later, they increased this by another £480,000 to allow the construction of the Bath extension to begin. A contract was let with T. & C. Walker and work began in 1872. Having released the

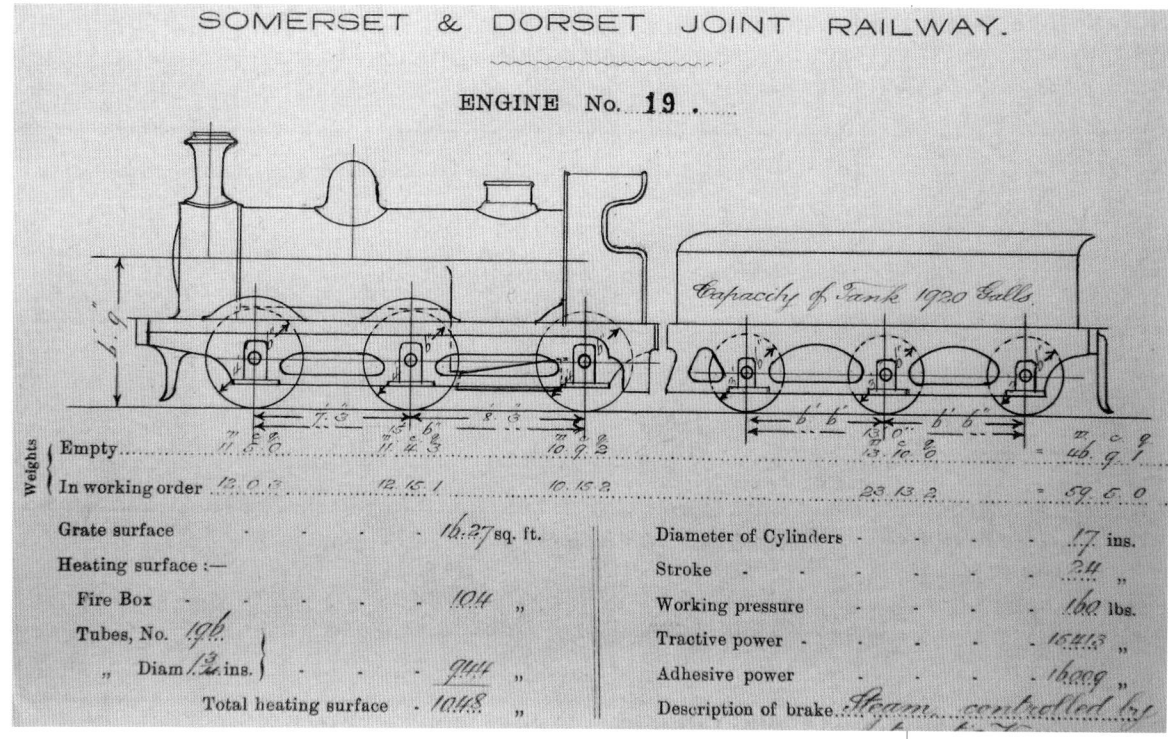

**The six** new 0-6-0s were given the numbers 19 to 24 when delivered in 1874. This meant that the engines with these existing allocations were either re-numbered or withdrawn and sold. Old numbers 1, 3, 4 and 5 were acquired by Fox, Walker & Co of Bristol as make weight in a deal to buy nine new tank engines. The drawing and photograph above and to the left were prepared and taken in the 1890s as part of the records kept by Locomotive Department. Each cost £3,190 and all but the last were in service when the line to Bath opened on 20 July. (Author)

**By 1874** Fox, Walker and Co were a well-established company in Bristol building significant numbers of small tank engines. They advertised widely and their engines were featured in magazines of the period (as shown here), often with flattering descriptions of their capabilities. It is little wonder that the company came to Slessor and Fisher's notice. (Author)

purse strings for the extension, the directors could hardly have refused Slessor's demand for more engines. With the new line due to open in 1874, a contract with John Fowler & Co of Leeds was agreed and the six 0-6-0 tender engines, specified by the Resident Engineer, were built with delivery set for the following year – one a month from March to August.

These six engines were virtually identical, except the last two had one modification. Fisher requested that they be fitted with taller domes above the middle ring of the boiler. He was concerned that, when a locomotive was hauling its load up a steep slope, 'priming' would occur, allowing water to spill over into the steam delivery system. The taller dome would make this less likely. Why only two engines were modified in this way isn't known or, for that matter, whether the others underwent this modification at a later date. Either way, it was a small detail and one unlikely to undermine the general impression that the S&D now had locomotives more suited to the demands of the line. If success is measured in years of service and longevity, then these proved to be valuable servants of the line, only being withdrawn in 1927/28.

Although now equipped with these powerful new engines, neither Slessor nor Fisher had forgotten the requirement, identified in the report of 1870, for new banking engines to assist trains over the hills on the Bath extension. This would meet the crucial condition of allowing heavier loads to traverse the line and, in theory, increase profitability. Here they turned to Fox, Walker and Co at their Atlas Works in Bristol to find a suitable engine. The company had only been formed ten years before in 1864 but had quickly produced nearly 400 locomotives. One area in which they had begun to specialise was in engines capable of working steep gradients, as their publicity material made clear. Whether Fisher looked at other types of locomotive available across the industry or sought expressions of interest from other companies isn't clear. All we seem to know is that in February 1874, three saddle tank 0-6-0s were ordered from the Bristol company at a cost of £2,350 each. With the S&D financially stretched by the construction of the Bath line, the deal saw the George England built locomotives 1, 3, 4 and 5 sold to Fox, Walker in part exchange; with a value of £2,550 for all four. This negotiation then allowed the S&D to order two more in August 1874 for delivery in February the following year. It soon became clear that more might be required as and when the Locomotive Superintendent could persuade the directors to open the purse strings again or find another procurement strategy.

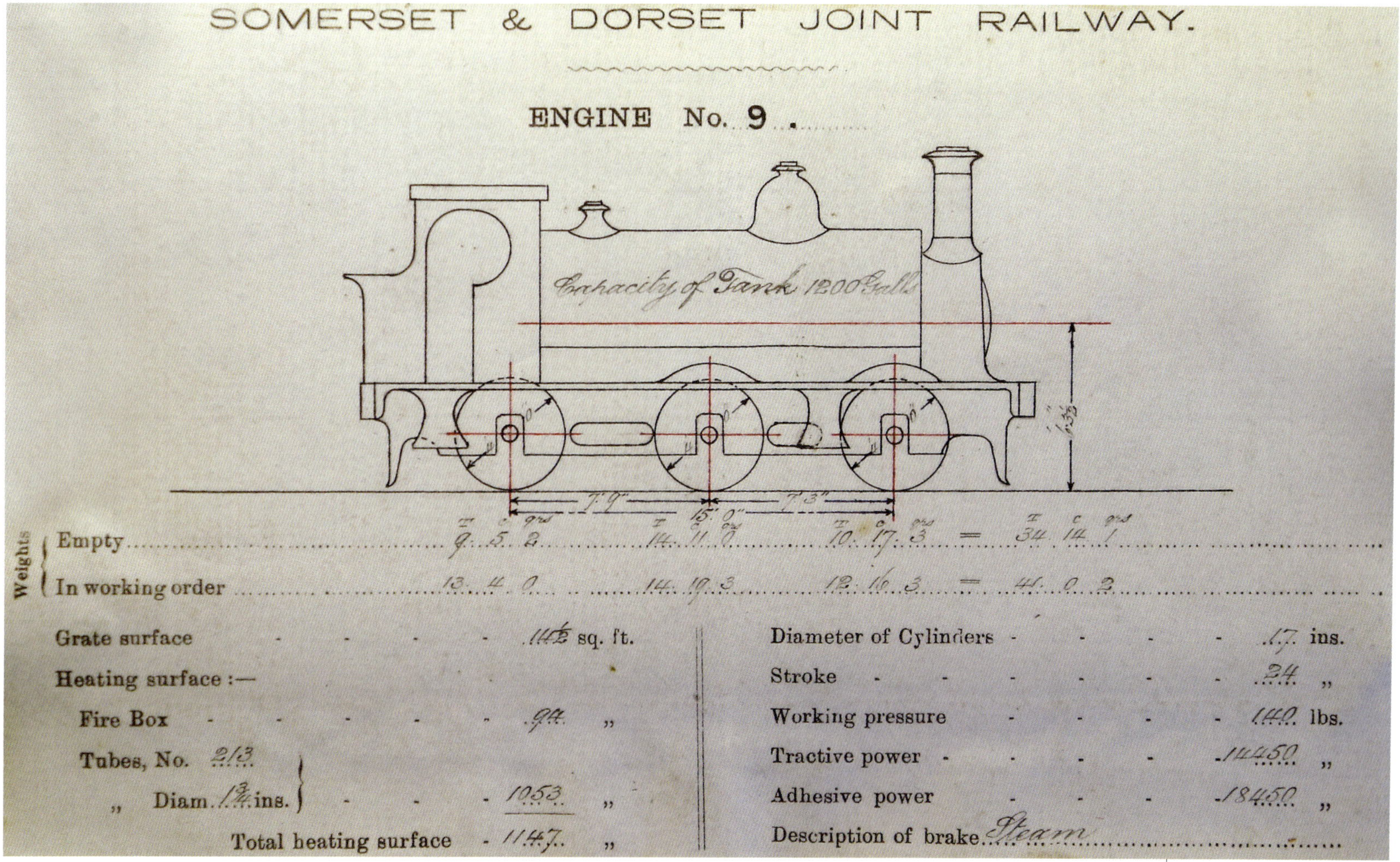

# SOMERSET & DORSET JOINT RAILWAY.

## ENGINE No. 9.

Capacity of Tank 1200 Galls

| Weights | | | | |
|---|---|---|---|---|
| Empty | 9. 5. 2 | 14. 11. 0 | 10. 17. 3 | = 34. 14. 1 |
| In working order | 13. 4. 0 | 14. 19. 3 | 12. 16. 3 | = 41. 0. 2 |

| | | | |
|---|---|---|---|
| Grate surface | 14½ sq. ft. | Diameter of Cylinders | 17 ins. |
| Heating surface :— | | Stroke | 24 " |
| Fire Box | 94 " | Working pressure | 140 lbs. |
| Tubes, No. 213 | | Tractive power | 14450 " |
| " Diam. 1¾ ins. | 1053 " | Adhesive power | 18450 " |
| Total heating surface | 1147 " | Description of brake | Steam |

*Above, left and overleaf:* **The words** strong and rugged apply to the nine saddle tank engines Fox, Walker and Co supplied to the S&D between 1874 and 1876, as the diagram above and the three photos on this and the next page make clear. They also seem to have been very reliable locomotives and served the line effectively until withdrawn from service between 1928 and 1934. Over the course of their lives, a number of modifications were undertaken. For example, No. 1 was adapted to become an 0-6-0 tender engine between 1888 and 1908 and then was reconverted to become a saddle tank once more, staying that way until 1930. No. 3 was fitted with an extended saddle tank. In 1889, No. 8 became a side tank 0-6-0 engine with 4ft 6in diameter wheels and remained so until 1908 when rebuilt with a Deeley boiler to become an 0-6-0 tender engine until withdrawn in 1928. (Author)

Once in service, these engines proved themselves more than capable of banking trains from Bath up the difficult incline through Devonshire Tunnel then on to Combe Down Tunnel and from Radstock to Masbury Summit. When not working these turns, they found ready employment as shunting engines, for which they were ideally suited. Such was the demand for these engines that, in 1876, four more were ordered, but with only limited funds to play with some creative thinking and hard negotiations were necessary. The only way the deal could be closed, without denuding the railway of much needed motive power, was to sell all six of the new John Fowler engines plus two of the Fox, Walker saddle tanks to a third party. In this case it was to a C. Christian of Bristol with whom a hire purchase agreement was reached. Records do not reveal who he was or his background. So one can only speculate on the reasons for his interest. He may have had a close association with the S&D or was simply a local entrepreneur with cash to spare. Either way, his intervention proved crucial and these eight modern engines continued to serve on the line. But this wasn't the full extent of the deal to acquire four more 0-6-0STs. Three surviving, and increasingly maligned, George England 2-4-0s, Nos. 2, 6 and 7, were sold to the Railway Rolling Stock Company of Birmingham and then hired back.

The total cost of the transactions with Christian and the Midland company would be £224 per month.

Despite these economies, the railway was again severely overstretched. The directors' ambitious plan to build the Bath extension had proved their undoing. The crippling debts, and the interest this generated, were more than they could service from the revenue earned. To survive any longer, they badly needed an injection of cash or new collaborators to share the burden. As 1875 dawned, a purchaser or partner was sought, with neighbouring railway companies being the most likely candidates. First up was the Great Western, but they stuttered or prevaricated over an offer perhaps due to the demise of broad gauge over the S&D and the cost of its reinstatement. This allowed the LSWR and Midland Railway to put together a joint bid which was accepted and on 1 November that year they took over responsibility for the line, despite strong opposition from the GWR. In this joint agreement, the Midland took sole responsibility for the locomotives and rolling stock and the LSWR the infrastructure.

So ended the first part of the S&D's life, a period that came to be defined by a constant battle for survival and the effect this had upon the company's ability to operate. While the Somerset Central and Dorset Central existed in isolation from each other, they relied upon the Bristol and Exeter and the LSWR for their locomotives and rolling stock. These well-established and secure companies were able to build and maintain engines of some quality for the age. So their new neighbours, with only limited services to support, could draw upon a fleet of some quality. However, when they amalgamated, this safety net was slowly pulled away and they had to look to their own resources when it came to engines, carriages and trucks. It became a battle for survival in which the company failed to buy the locomotives with the necessary power it needed. So they settled for George England engines and then bought others because they were available. A cautious approach in the circumstances, but not one likely to achieve the desired results; cobbled together solutions are rarely successful and only in 1874 did locomotives begin to arrive that were truly suited to the demands of this singular line. Then a fiscal reality descended, and the company's spluttering ambitions came to an end. One thing is certain, though – with the Midland Railway taking responsibility for motive power, a tougher, a more professional group of men would soon begin to exert their influence over locomotive design. With them would come new ideas and new ways of working that would transform the fleet and the way the railway was run.

## Chapter 3
# THE MIDLAND YEARS PART ONE (1876 to 1903)

Like the S&D, the Midland Railway began in a modest way during the early 1830s, but unlike the S&D it had major industrial customers to sustain its growth – the coalfields of Nottinghamshire and Leicestershire. And nearby there were rapidly expanding towns and cities, plus many smaller communities, all of which could support a fledgling company as it struggled for survival. With such a market emerging, entrepreneurs were soon attracted by the possibilities which railways presented. This coincided with the boom in rail transport across the country, itself supported by rapidly emerging technology, particularly in the field of locomotive design. Soon, businessmen were developing investment schemes and seeking the specialist advice of the few men then becoming expert in this field of engineering.

William Stenson, a Leicester man, who had begun development of coal mines in the virgin land around Whitwick in the 1820s, saw his business interests expand rapidly. In seeking to move the huge volumes of coal his mines were producing, he soon realised that horse drawn carts were limiting growth and profits. In looking for solutions to this problem, he visited the Stockton and Darlington Railway in 1827 and conceived the idea of a line in his area. He then developed this proposal with fellow businessman, John Ellis, who was described as a Quaker weaver. Through him, George and Robert Stephenson were drawn into this emerging enterprise and soon the idea of the sixteen-mile-long Leicester to Swannington Railway was born. In May 1830, their ideas were given legal status by an Act of Incorporation. Robert Stephenson was appointed project engineer and supported by £90,000 from investors, including £2,500 from his father, soon began work, employing several contractors, including Daniel Jowett, along the way. The first service began in July 1832, but the entire line didn't open for traffic until November the following year when the section between Long Lane and Swannington was completed.

With this first step, the creation of the Midland Railway was completed. However, it would be another eleven years before it came into existence, with many twists and turns along the way. New companies would emerge, created by other equally ambitious men across the region. Then, driven by the needs of competition, expansion and capitalisation, there followed takeovers and mergers. In this way, the Midland Counties Railway, opened in 1839, of which the Leicester and Swannington became part, then joined with North Midland Railway to form the Midland Railway. Two years later, they were affiliated with the broad gauge Birmingham and Gloucester Junction Railway, whilst the GWR dallied too long over its acquisition, as they did with the S&D thirty-five years later. And with this, the MR created a network with its hub at Derby. In time, its limbs would stretch across the Midlands, touching Sheffield, Leeds, Carlisle in the North, Lincoln and Peterborough to the East, London

to the South and Bristol to the South West.

As the railway grew, so did the workshops at Derby and by the time the company took over responsibility for the S & D's locomotives and rolling stock, it was one of the major centres of railway activity in Britain. Over this organisation, five Locomotive Superintendents, then Chief Mechanical Engineers, would rule for a period of seventy-eight years as though they occupied personal fiefdoms. In an age of subservience, with few civil liberties, limited employment rights and minimal health and safety regulation, such an outcome was inevitable, and the Works at Derby were no different to the mass of other employees at this time. However, each man's approach varied according to their personality and peccadilloes.

It was they who determined locomotive policy, decided strategy and drove through design. And it was they who would determine the type and level of support the S&D would receive. First of them, Matthew Kirtley, was the longest serving, and died in harness in 1873. After twenty-nine years occupying the top seat, his influence over the company was immense and his ideas would affect the way Samuel Johnson, his successor, worked for a while. On paper, Kirtley had the least to do with the S&D.

**During the** years that the Midland Railway controlled locomotive matters on the S&DJR there were four Locomotive Superintendents (the post was re-titled Chief Mechanical Engineer in 1909). And before that there was the dominating presence of Matthew Kirtley (top left) who directed business from 1844 to 1873 and was a major influence on the company's design and production programme. Following him came Samuel Waite Johnson (1873-1903) (top middle), Richard Deeley (1904-09) (top right), Henry Fowler (1909-15 and 1919-22) (lower left) and James Anderson (1915-19) (lower right). Each brought influence to bear on the design and operation of locomotives that ran on the S&D, but in varying degrees. Anderson probably had least time to make his mark only being in temporary charge, as Deputy CME, whilst Fowler spent time away on war related duties with the Ministry of Munitions. However, before that he was Works Manager and Chief Locomotive Draughtsman, in which roles he exerted considerable influence over motive power policy. (Author)

**Derby Station** and Works as it appeared in the late 1880s when at the centre of all activity concerning the production and supply of locomotives to the Somerset and Dorset Joint Railway. It retained its status in Britain's railway industry through the life of the LMS into BR days. (Author)

Nevertheless, in championing and developing best practice through good design, standardisation and rebuilding existing engines where possible, he provided an excellent example for others to follow.

When it came to design, he believed in the benefit of having inside cylinders and for many years followed George Stephenson's ideas on frames. As a result, he used a wood and iron sandwich variety until, in 1852, he began experimenting with outside plate and double frames, with horns connected by rectangular tie bars. This solution became a feature of his work. Over the years, he was responsible for building or rebuilding many hundreds of locomotives which were noted for their good steaming qualities, strength, reliability and longevity. Of particular note were his Class 1 2-4-0 passenger tender engines, Class 1 0-4-4 passenger tank engines, Class 1 0-6-0 goods tender engines and Class 2 0-6-0 goods locomotives. Construction of two of these classes was continued by Johnson and substantial numbers of each passed into the service of the London, Midland Scottish when formed in 1923. An excellent epitaph on the success of his ideas and his designs.

Kirtley was described by Hamilton Elllis in his book *The Midland Railway*, as a 'rough, jovial, homespun sort of engineer who had once been a fireman. His engines were like him: rugged, not conspicuously elegant, tremendously substantial and lasting'.

Samuel Johnson was quite a different kettle of fish. Ellis picked

**This family** group contains the two men who would direct the production of locomotives and rolling stock on the Midland Railway. The gentleman to the left is Thomas Clayton who became Carriage and Wagon Superintendent at Derby in 1873. Up to this time the task had rested on Kirtley's shoulders. To his right is Johnson. (D. Neal/Author)

out the main ingredients of his personality as 'a pillar of the church, a man of kindly severity and meticulous habit. His work, like himself, was meticulous. A Johnson locomotive was designed like a work of art and made like a watch. Mechanical judgement was the virtue. The Johnson engines were amongst the most beautiful…. In finish they excelled those of any other railway in the world'. His professional background reflected the way the railways had grown and the many refinements made in training and design since the days of Kirtley's baptism in the industry.

At 13, Kirtley joined the Stockton and Darlington Railway and progressed to become a fireman then driver on the London and Birmingham, before promotion to foreman with the Birmingham and Derby. So rapid was his progress that he became their Locomotive Superintendent in 1841 and the MR's three years later. He learnt on the job, so to speak, there being few if any railway apprenticeship schemes at the time to follow. Johnson was born near Leeds in 1831 and served an apprenticeship at the Railway Foundry in Leeds, then run by Fenton, Craven and Co but becoming E.B. Wilson early in 1847. Whilst there, he appears to have spent time in the drawing office acquiring these essential design skills. From here, he

**Johnson's time** with the Great Eastern proved to be a creative one in which the ideas he had developed over twenty or so years with various companies found a voice. His first design of note during this period was his No. 1 Class 2-4-0, which appeared in 1867 (nicknamed 'Little Sharpies'). The prototype is pictured here in 1906 after undergoing modification by James Holden, then Loco Superintendent, and his team. Some members of this forty strong class lasted in service until 1913 suggesting that the design was considered a sound one. (D. Neal)

progressed to the Great Northern where he worked and studied under Archibald Sturrock. 1859 found him working in the Gorton Works of the Manchester, Sheffield and Lincolnshire Railway as Works Manager under Charles Sacre, who joined the company on 1 April 1859. Johnson remained at Gorton for five years and would have observed and assisted in the design work of this talented engineer.

Clearly an ambitious man, he then applied for the post of Locomotive Superintendent with the Edinburgh and Glasgow Railway. Here his design credentials were further improved by his work on a number of double-framed 2-4-0 passenger locomotives. But in 1865, this company amalgamated with the North British Railway and in the transition period that followed he decided to leave. As a result, he joined the Great Eastern, replacing Robert Sinclair as Locomotive Superintendent at its Stratford Works in London in 1866. Here he remained until the Midland Railway beckoned seven years later.

These years proved to be fruitful ones. He soon discovered that the GER was in urgent need of an injection of new passenger and goods engines to reinvigorate its tired, and undoubtedly expensive to maintain, fleet. His first move was to bring in Neilson, Reid & Co 2-4-0 mixed traffic engines in 1867 as a quick fix, then built three more similar engines at Stratford. These became prototypes used to develop the No. 1 Class of 2-4-0s, forty of which were built by 1872. This was followed by sixty 417 Class 0-6-0s built between 1867 and 1869 and then fifty of the more powerful 477 Class 0-6-0s. In due course, he added the Class T7 0-4-2s, the 200 Class 4-4-0Ts, the 209 Class 0-4-0 and the 134 Class 0-4-4T to the list of his achievements. And so the programme ran on until his departure to the MR in July 1873, by which time he had begun to investigate designs that employed leading bogies. This exploration began with his Class C8 4-4-0 express engines. In fact, only two were built and these in 1874 when William Adams was Locomotive Superintendent. By this time, Johnson was well ensconced at Derby where he was in the process of taking his ideas to new levels in a programme that would affect the S&D and its ability to become a successful business, something that had so far escaped this railway.

Once the leasing agreement was signed on 1 November 1875, and the Midland took over responsibility for the locomotives and rolling stock and the LSWR the railway's infrastructure,

**Johnson's 417** and 447 Class 0-6-0s appeared between 1867 and 1875 by which time 60 and 50, respectively, were in service. They underwent some modification work, but all had shorter lives than his No.1s. The last surviving 417s disappeared in 1889 and the 447s in 1902. Engine No. 420 is pictured here. (D. Neal)

**Johnson appears** to have been eager to explore the concept of engines with leading bogies. The result was his two very elegant C8 4-4-0 passenger express locomotives which appeared in 1874 a year or so after he had left the GER. The C8s are believed to be the first 4-4-0 engines in England to be fitted with inside frames and inside cylinders. It was a design he would continue to develop with the MR. (R. Hillier)

representatives from each company undertook detailed surveys to establish the extent of the problems they had inherited. The results were not edifying and revealed many shortcomings. Today, the railway would be described as not fit for purpose so serious were the problems uncovered. For example, of 204 passenger trains observed, only 17 were worked to time and there were five engine failures recorded. Goods services fared little better and didn't run a single train to time. Safety wise, three derailments were also observed during this period, which, luckily, were fairly minor in nature and didn't result in any casualties.

It seems likely that S&D staff were asked to comment on issues raised by the inspecting team, which is normal business practice in such circumstances. If so, Slessor and Fisher, who was soon to be appointed Resident Locomotive Superintendent, would have pointed to the unsuitability of many engines for main line duties and the shortage of trained staff. These were two issues well known before the Midland team arrived. They would also have pointed to the extent of single track sections of line and the delays to service this could cause. Then there were issues over exchange facilities at Templecombe which called for reversal of all trains using the main platforms. These were all valid points which the MR's team probably accepted. However, they would have considered other causes and may have had concerns over laissez faire attitudes amongst the staff and the standard of locomotive maintenance with five failures observed in such a short time.

Although the MR had the expertise and industrial muscle to make changes and improve the service immediately, they were limited in what they could do by the terms of the lease. The loan of some locomotives was possible, and took place, but no major expenditure on new equipment and no significant changes to working practice until an Act of Parliament confirmed that the new arrangements were sanctioned. These terms and conditions came into force on 13 July 1876, too late to allow any significant changes to be made that might have helped avert the rail disaster at Radstock

**The aftermath** of the Radstock accident on 7 August 1876. This comparatively peaceful scene belied the violence of the accident and its aftermath with thirteen dead and thirty-four injured. Incompetence, lack of discipline and an inability to follow regulations were the primary causes. Individuals were at fault but in many ways their behaviour mirrored the parlous state of the railway on many levels. (D. Neal)

**One small** example of the Midland Railway's industrial muscle on display. In 1873 the MR could muster more than 1,000 locomotives and this had risen to more than 2,300 by 1897. In 1876, by comparison, the S&D's numbers were tiny, with less than thirty on their books of mixed quality and availability. This picture shows Erecting Shop No. 2 at Derby in the 1880s with a mixed group of engines undergoing maintenance. In the foreground is a 1357 Class 0-6-0 designed by Johnson and team, one of 110 built between 1878 and 1884. Another identifiable locomotive is No. 939, a Kirtley designed 760 Class 0-6-0 engine, one of ten built by Kitson and Co out of a total of 316 ordered by the Midland Railway. In the background sits another 0-6-0 No. 1220, one of 120 Class 1142s built between 1875/76. (Author)

on 7 August, and save thirteen lives and much misery in the process. It was an incident that perfectly captured the S&D's failings in a most graphic way, leaving no one in any doubt that radical action was essential.

The arrival of both the MR and LSWR was a godsend for the struggling company, but so was the recruitment of Robert Dykes as Traffic Superintendent and Alfred Colson as Resident Engineer that year. With Robert Read as General Manager and Fisher as Locomotive Superintendent, until his death at Burnham on 28 April 1883 when only 46 years of age, the company had four men in place with the drive and ambition to make changes and seek the resources necessary to improve services. Nevertheless, progress in these situations can be slow and is often incremental in nature relying, as it does, on funds being made available, which as we have seen wasn't always possible on this cash strapped line. Meanwhile, at Derby, Johnson, unhindered by such a restricted budget, was soon seeking to improve the S&D's locomotive fleet, by advancing his ideas on design in the bigger world of the Midland Railway.

Initially, he concentrated on developing an 0-4-4 tank engine for the Midland, building on experience gained during his

**Johnson produced** a number of 0-4-4T engines for the Great Eastern (thirty, with the first appearing in 1872) and then the Midland Railway. Between 1875 and 1903 155 were constructed at Derby, by Neilson & Co or Dübs. These elegant, well balanced locomotives are recorded as having been a success when performing main line passenger duties on the S&D and the MR. The engine pictured, No.1636, was one of the sixty-five 1532 Class which was built at Derby between 1881 and 1886. During his tenure, Johnson led in changing the colour of engines from bluish green to crimson to match the MR's carriages. 1636 was the last locomotive to receive a coat of green. These 0-4-4Ts had long lives, with some examples lasting in to the late 1950s. (Author)

years with the GER. These new engines began appearing in 1875 and would eventually rise to 155 in number. Thirteen of them would be assigned to the S&D between 1877 and 1885 and would play a key role in the line's history. He then began producing a modified version of Kirtley's 890 Class 2-4-0 express engines but chose to use 6ft 2½in coupled wheels, mixed frames and boilers with 1,225sq ft of heating surface producing 140lb. of pressure. These appeared in 1876 and were quickly followed by other versions with coupled wheels ranging in size from 6ft 6in. to 7ft.

At the same time, his work with 4-4-0 tender engines continued with ten being built by Kitsons along the lines of his two GER prototypes, but with slightly larger cylinders and 6ft 6in wheels. These were built specifically to work over the Derby to Manchester line, where the curves and gradients through the Peak District were severe. Six of them passed into LMS ownership and the last engine was withdrawn from service in 1930, which supports the view that the design was a successful one. During 1877, a similar version with 7ft coupled wheels was added to the inventory, this time built by Dübs and Co of Glasgow, a forerunner of the North British Locomotive Company. There was then a pause of five years before the next 4-4-0s were built, but from then other versions gradually appeared. There were some variations in wheel size, wheelbase, boiler pressure and cylinder sizes, but generally speaking, it was considered that there was a degree of standardisation in their design. Over the years, the Johnson 4-4-0s and their successors would become a common feature of life on the S&D, with later versions built at Derby lasting into BR days. All in all, they played an important role in this line's history as we shall see.

There were two other classes that Johnson developed which would also have an impact on the S&D. First came his 0-6-0 goods locomotives introduced in 1878. The John Fowler built 0-6-0s, acquired with the intervention of Mr Christian in 1874, had by the strength of their performance, already paved the way for more engines of this type. Such strong, reliable locomotives could manage the challenges this line presented, particularly over its more difficult sections, especially the steep inclines found on the Templecombe Junction to Bath portion. Their success created a demand for more of this type. Johnson obliged by allocating engines from the MR's massive construction programme that would see 935 0-6-0 tender engines built between 1875 and 1908 at Derby and by a number of contractors. They all followed a very similar pattern, so might be seen as a single class, but there were variations which allowed them to be tailored to meet specific needs.

The final group of engines provided by Johnson were five with side tanks required for shunting duties at the Radstock collieries. Though lacking the prestige of the bigger engines that

**The gradual** evolution and refinement of Samuel Johnson's 4-4-0 express passenger engines. Top – Engine No. 1320 with 6ft 6in coupled wheels built by Kitsons in 1876. Middle – No. 1756 built at Derby in 1886. Bottom – No. 819 that appeared in 1902 with a bogie tender. This was one of Johnson's last designs, but he did add an experimental three-cylinder compound version to the fleet in the same year, with two sets of reversing gear for low and high pressure systems. (D. Neal/Author)

**A typical example** of Johnson's work with 0-4-0ST engines while with the Midland Railway and the influence this had on the selection of similar engines to support the mines around Radstock. In this case, the engine is one of ten 1116A Class locos built at Derby in two groups of five during 1893 and 1897. It is pictured here in LMS colours with a 4-4-0 tender engine next in line; a type that would itself become a familiar sight on the S&D, though not this particular engine, apparently. (D. Neal)

populated the line, their role was just as important in many ways. Up to 1882, mine owners had been happy to use horses to pull coal trucks, but such a practice seemed increasingly inefficient and out of step with the rapidly modernising world. The Midland Railway, with its origins in the Nottinghamshire and Leicestershire coal industry and a massive trade in pulling coal trains, had some experience in this field to guide their locomotive policy. In due course, this led to the appearance of four 0-4-0STs and a single 0-4-2ST to support the Radstock collieries. At the same time, Johnson and his team were considering the wider need for such engines and established that a sizeable demand existed across the MR. This led to the production of three types of 0-4-0Ts between 1883 and 1903; the 1322, 1116A and 1134 Classes, ten of each being built at Derby.

So, this is the extent of Johnson's development work as it effected the S&D. It has to be said that there was a depth and variety to his ideas which encouraged flexibility in design to match a variety of conditions. It was this quality that the S&D's small fleet of locomotives needed more than anything else. However, there were other engines he developed that didn't have a direct impact on the Joint Railway. For example, one great triumph was his graceful 'Singles' – 4-2-2 passenger express engines built between 1887 and 1893. Sadly, these don't appear to have ventured onto the S&D's metals at any stage so do not feature in this story.

Having established the strengths and weaknesses of the S&D's locomotive and rolling stock fleets in the review of 1875, Johnson and his team would have been in no doubt about the problems they faced. Slessor and Fisher's efforts had paved the way and with the arrival of the six 0-6-0 tender engines in 1874 and nine 0-6-0STs by August 1876, a good start had been made. But more was required and in the short-term, Johnson temporarily boosted the Joint Railway's fleet with the addition of six MR 0-6-0 and 0-4-4 engines. This gave them some breathing space but no more. A longer-term solution was essential and this involved careful review, the preparation of sound specifications and a procurement programme approved by the S&D's joint committee. Even though leasing fees brought

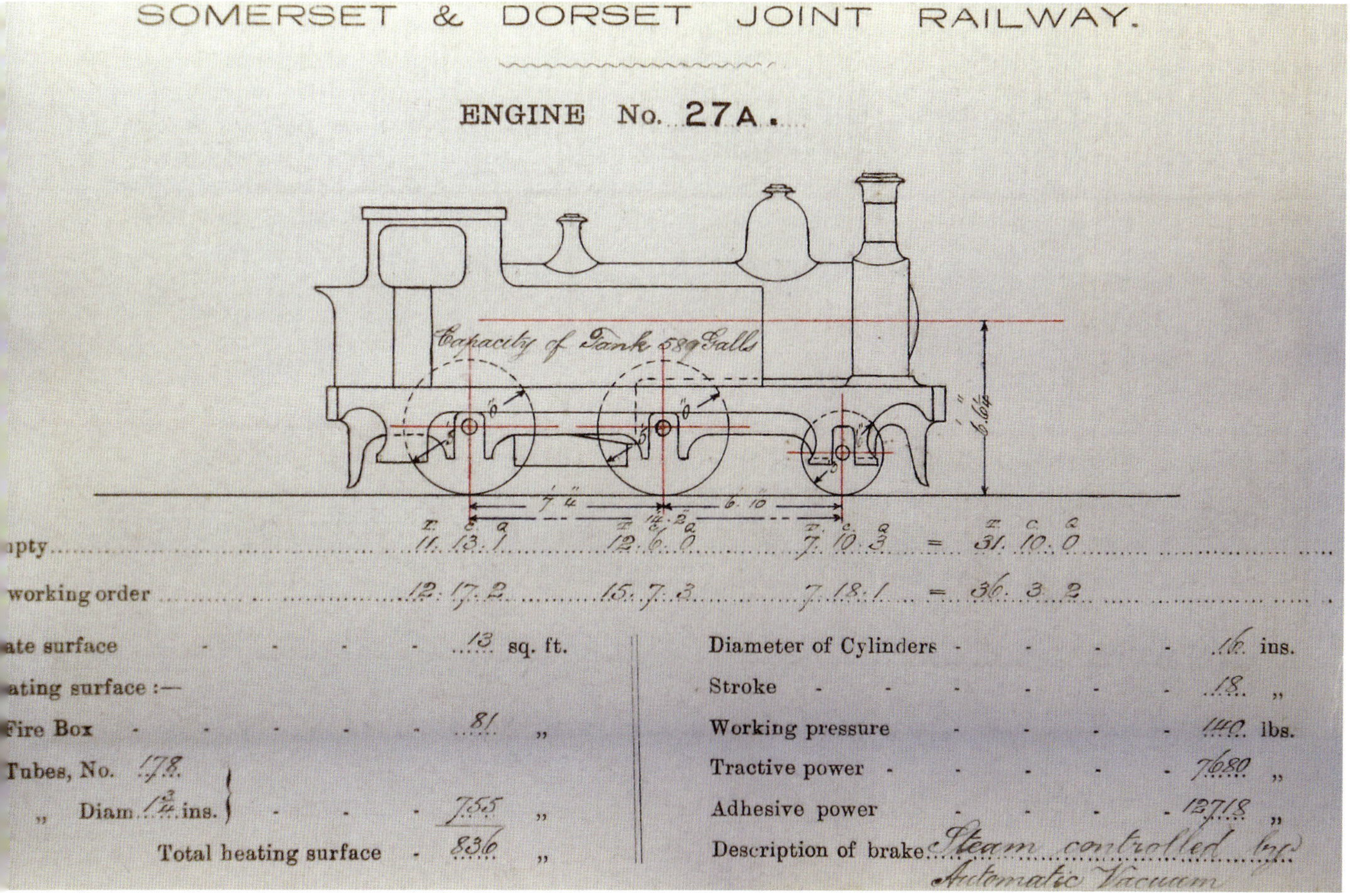

**The review** conducted by the MR before the launch of the S&DJR in 1876 was critical of the motive power accumulated by the old company, particularly the George England built 2-4-0 tender engines. Some were sold to the two new leasing companies, but two others, Nos. 6 and 7, were rebuilt as tank engines in which state they continued to serve the line until 1889 and 1925 respectively as 26/26A and 27/27A. If longevity is a measure of success, 27A was remarkably successful in lasting sixty-six years. However, its long service suggests something more – that it was a locomotive that had proved its worth. The engine is portrayed here as it appeared at the beginning of the nineteenth century. (Author)

some stability to its finances, any lavish spending was unlikely to be approved.

In this situation, ordering new locomotives meant selling some of the old to help balance the books; in this case six of the George England 2-4-0s, Nos. 9, 10, 12, 13, 14 and 15, with three of each being purchased by the MR and LSWR in 1878. The impression given is that their departure was not mourned by engine crew, who had struggled to make these underpowered engines a going concern. Their sale value was unlikely to be high and their value to the two leasing companies probably minimal, but the transaction was one of the few ways the S&D had to finance such a programme. However, the England 2-4-0s didn't completely disappear from the line and four of the 1861 group soldiered on for a few more years. Number 2 remained pretty much unaltered and acquired the soubriquet 'Old Number 2' in the process, although becoming No. 12 in 1876. In 1884 and 1888, Nos. 6 and 7 respectively would be rebuilt as 2-4-0 tank engines, while No. 8, which was converted into a side tank engine in 1862, reverted to simply tank status in 1876. The changes didn't end here though. In 1883, it regained a side tank, but reverted again in 1904, in which condition it was scrapped twenty-four years later. No. 6 predeceased it in 1889 and No. 7 in 1925.

With the decks cleared, the scene was set for the first of Johnson's newly built engines to appear. In the circumstances, their impending arrival was probably underscored by a strong sense of expectation, not least by the men who had to crew them. If so, the look and

**The LSWR** bought three of the S&D George England built locomotives in 1878 and they continued in use for a time. No. 12 became No. 148 *Colne* and is captured in this faded sepia picture three or so years later. After repair at Nine Elms, it was employed on light duties in the Dorchester-Bournemouth area initially and was then rebuilt in 1881 with new cylinders, copper firebox, a new cab with side windows and stovepipe chimney. In 1883, it was assigned to the engineer's department and became No. 3, being re-christened *Stephenson* along the way. The engine was withdrawn in December 1889 having achieved 508,964 miles. Its boiler was salvaged and continued in service in static form providing heating to workshops. (D. Neal)

**Another George** England engine to survive on the S&D was the 2-4-0 built as a tank engine and given the number 8. However, it underwent a number of conversions during its lifetime, becoming a saddle tank in 1862 and then a tender engine in 1876. But it was nothing if not flexible and in January 1883 it became a saddle tank again, as shown here in the upper photograph, and finally (lower picture) an engine with side tanks once more having been re-numbered 28/28A in the process. In this state it remained until withdrawn from service in April 1928 after sixty-seven years of life. (Author)

**As demonstrated** by this brochure, the Avonside Engine Company was well-established by the time they began building nine 0-4-4Ts for the S&D to a Johnson design. As Britain's main line locomotives grew larger and more powerful, they found a niche market in building smaller, specialist engines. They survived until 1934 when lack of work and financial problems led to liquidation. The following year their designs were acquired by the Hunslet Engine Company of Leeds. (Author)

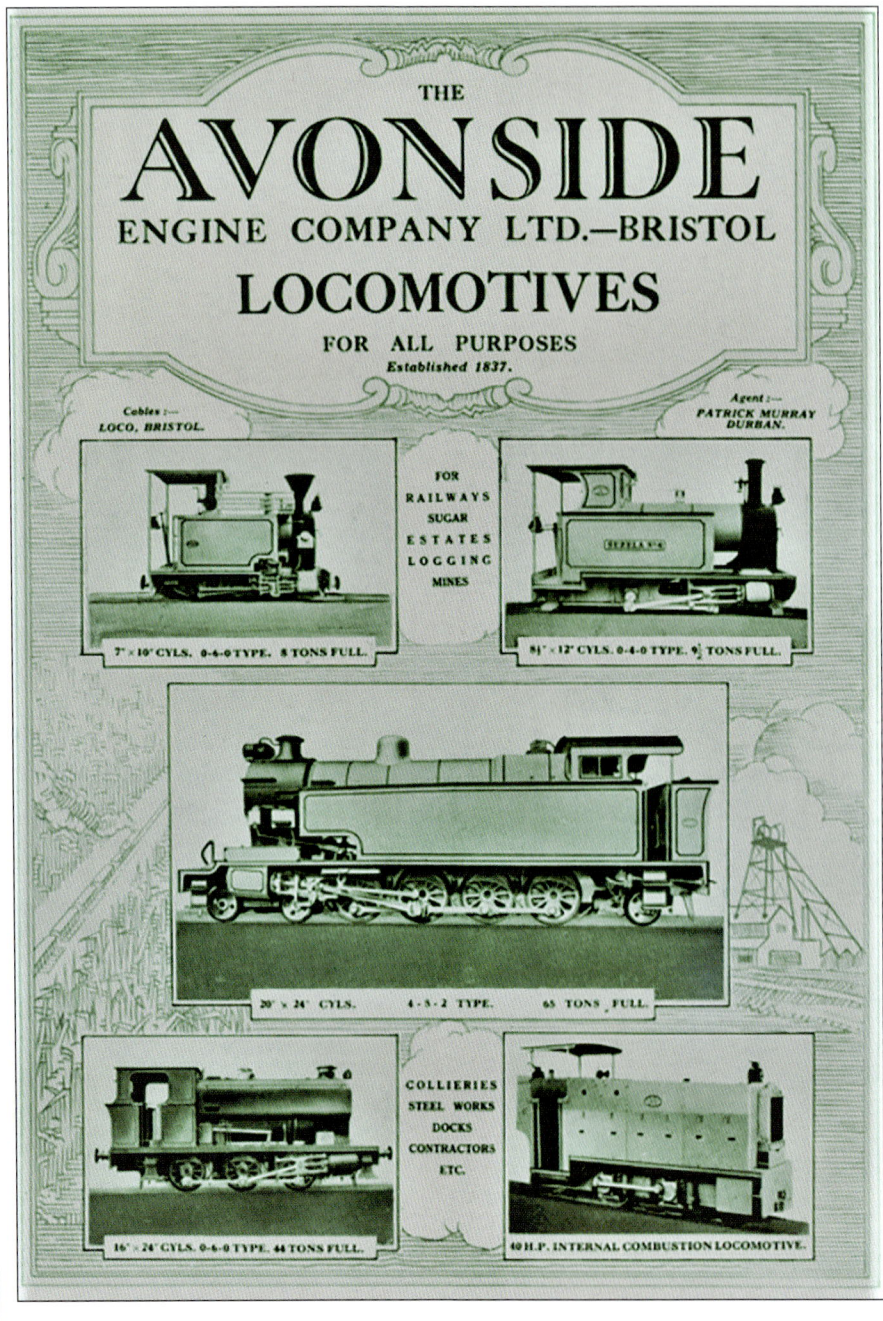

performances of the early arrivals wouldn't have disappointed them, though their cabs might have. For reasons that are not entirely clear, Johnson had not yet grasped the need to give footplate men greater protection from the elements, and simply provided weather boards.

First up were nine Johnson designed 4-4-2 tank engines procured through a contract awarded to the Avonside Engine Company, then based in Avon Street, St Philips, Bristol, in 1877. This was a company that came into existence during 1837 under the name of its founder Henry Stothert. With the railway industry booming, it was a good time for business and they began producing locomotives, two, most famously for the GWR in 1840 for its newly inaugurated line from Bristol to Bath. In due course, the company was joined by Edward Slaughter, who from 1837 had been Brunel's assistant engineer, as managing partner to Stothert. By 1856, the company had evolved into Slaughter, Gruning and Co and under this name had bid to build eight 2-4-0 engines and six tenders for the Somerset Central Railway only to see the work go to George England in 1861. Despite this, they continued to grow, with orders for main line locomotives arriving from British railway companies and from overseas. During 1864, there was another change in name to the Avonside Engine Co Ltd and a further expansion of the business marked by an order from the GWR for twenty 2-4-0 Hawthorn Class locomotives. It was in this guise that they began producing the 0-4-4Ts for the S&D. For Avonside these engines were something new to build, but for Johnson and the MR they were a known and tested type by this stage.

Before the arrival of the Avonside built 0-4-4Ts, some concerns had been expressed by the Engineering Department about the use of front coupled tank engines for main line express passenger trains. This debate came to a head with the arrival of three Johnson 0-4-4 tanks on loan to the S&D. There were concerns over rough riding and possible damage to the track this might cause. However, Johnson felt no qualms about using them for such duties. Nevertheless, to calm these fears, he did seek improvements to their springing and bogie control. This work would also have been encouraged by a separate report by F.W. Webb and Patrick Stirling, acting as independent consultants, that found the permanent way to be of 'inferior quality'. Despite this, they concluded that it was sufficiently safe for these engines to work these passenger services if speeds were limited and the track underwent a daily inspection. These sorts of decisions can occasionally come back to haunt those who make them, especially when others have offered contrary advice. And on 26 February 1877 this nearly proved to be the case for Johnson who had already committed the company to buy the nine 0-4-4T from Avonside.

In the early morning darkness, a Midland 0-4-4T, No. 1262, left Wimborne with the first through passenger train of the day. When passing between Templecombe and Wincanton at 35mph, with all apparently running well, the leading wheels left the track and ran onto the sleepers for 200 yards or so. This violent movement then dragged the second set of coupled wheels off the rails and with this the locomotive turned

**The final** locomotive of the nine 0-4-4Ts supplied by Avonside in November/December 1877 at a cost of £1,960 each; a price considerably lower than the most expensive bid by £535. The numbering of these engines was split into two groups – 10 to 14 and 29 to 32. When this photo of number 32 was taken, a very basic, easily constructed cover had been fitted over the two weather boards 'protecting' the crew. Nevertheless, this was a minor issue when compared to the quality of these new engines. In 1882, engine No. 32 was involved in a head on collision with engine No. 8 in which one person was killed and fourteen injured. (D. Neal)

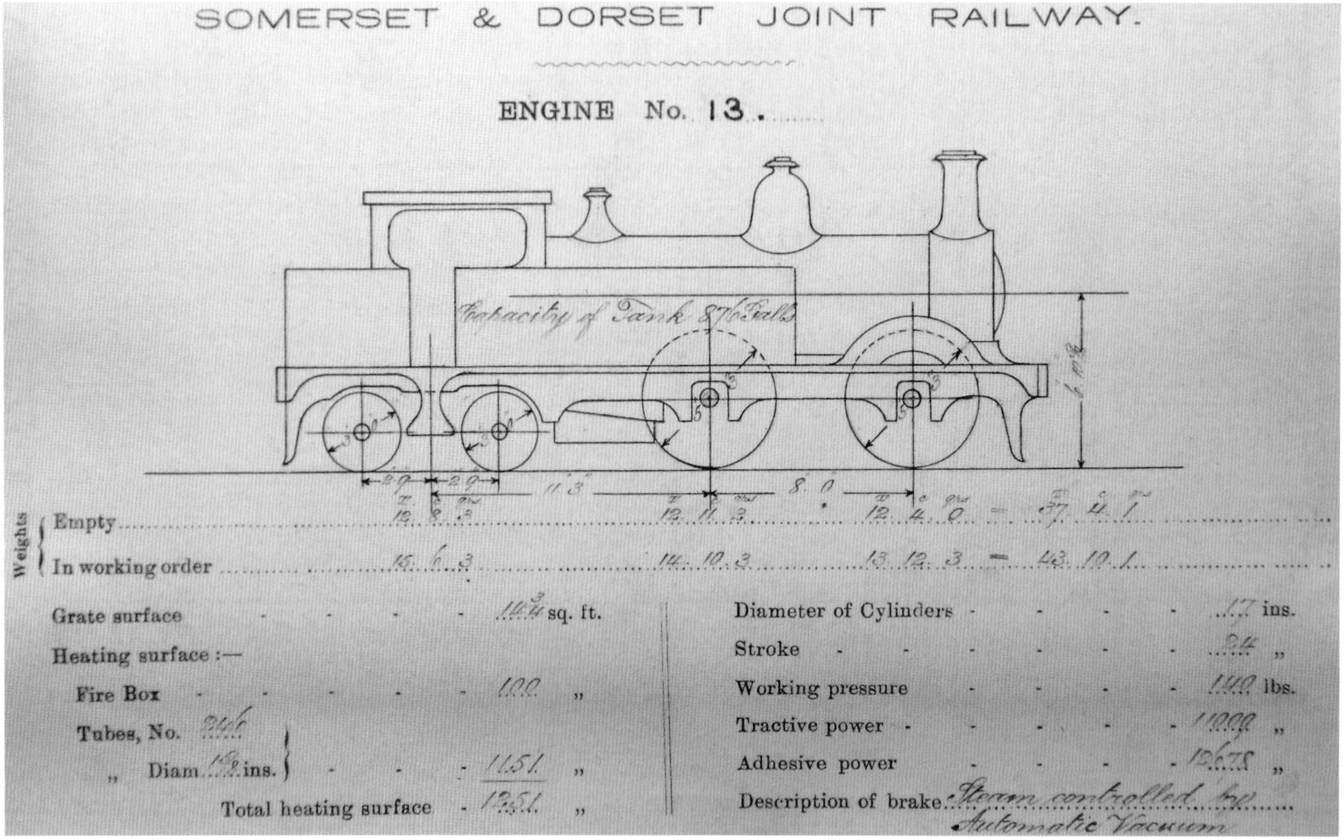

# SOMERSET & DORSET JOINT RAILWAY.
## ENGINE No. 13.

**Engine No.** 13, the fifth 0-4-4T supplied by Avonside in diagram form with basic dimensions added and (below) the reality. Though similar to other locomotives of the type built by the MR, they were slightly smaller, lighter versions. The bogie axles were more widely spaced, the fireboxes were 6in. shorter and the combined heating surface was 107sq ft by comparison to the 120sq ft of the MR engines. In this form they served the company well, with their arrival allowing engines borrowed from the leasing company to be returned. (Author)

**No. 11** photographed at Glastonbury on an unrecorded date. This engine was reboilered twice – in 1892 and 1909 – and was withdrawn in September 1930. (Author)

**A familiar** scene for many decades on the S&D. An unidentified Johnson 0-4-4T pauses at Blandford Station with its rake of coaches on a sunny, and judging by the straw boater on display, summer's day sometime towards the end of the nineteenth century. The signal box captured in this photo, on the up platform, was replaced in 1893 by a new larger box on the other platform, so this photo predates this changeover. (D. Neal)

**The dawn** rises and artists rushed to capture the scene of the crash on 26 February 1877 in the absence of cameras. In reality, they captured a stereotypical view of any crash suitably dramatized for readers seeking a sensationalised account of an accident. The carriages were reported as having remained upright, with only the the brake van seriously damaged. The engine being upside down and badly knocked about probably sustained a greater amount of damage, necessitating its return to Derby for repair. (D. Neal)

over burying itself in track ballast, pulling all the carriages off the track as it went. Driver Moorland was killed, while the fireman, guard and six passengers were injured. Word would quickly have passed to Johnson and one wonders if for a time he believed his engine policy may have contributed to the accident. Eventually, the Board of Trade inquiry exonerated him completely, finding the loco to have been in good working order and worked properly by its crew. Of far greater concern to Captain H.W. Tyler RE, the inspector, was the state of the track and the poor standard of inspection and reporting. If the latter had been better, it might have discovered damage inflicted on this section of track by other trains. In particular, he highlighted 'damage by the passing of earlier goods train or trains which had so weakened it that seven chairs and three sleepers failed to support the heavy bogie tank'.

In the aftermath of the crash, the S&D could not avoid investing heavily in track maintenance, replacement and better drainage. When 91½ miles had been raised to the required standard, nearly £135,000 had been spent on the work, but this is what happens in any business struggling to survive where preventative maintenance of machinery or infrastructure is sacrificed in favour of simply undertaking repairs when problems occur. Whilst this work was being undertaken, front coupled locomotives were restricted, the LSWR and MR providing ten 2-4-0s at different times to cover the shortfall. If these problems led to a re-think of his locomotive policy, Johnson clearly didn't feel that the accident undermined his plans and the first of the new 0-4-4Ts duly arrived in November that year, after a work up period in the Midlands.

Once in service, their qualities were soon apparent. The more widely spaced bogie axles, less weight than their MR cousins and better suspension, aided by bogies of the Adams sliding type regulated by rubber side springs, seemed to suit the S&D's renewed and improved permanent way. They quickly established themselves pulling expresses from Bath to Wimborne, but at this stage the trains they pulled were relatively

light and the schedule they had to keep quite generous. However, this gradually changed and in the early 1880s loads increased, timings were tightened up and through traffic from the North to Bournemouth increased, especially during the summer period. All this placed an extra strain on the 0-4-4Ts and could have presented difficulties to these or any other passenger tank locomotives over such a route. However, they appear to have coped with the work and sufficiently impressed the joint committee and Fisher to request four more. Johnson duly agreed with this proposal, set the tender process in motion and then awarded a contract to the Vulcan Foundry in April 1883 for these engines at a price of £2,295 each. For some reason, the delivery date that year couldn't be met, probably due to other production commitments. As a result, one engine appeared in December 1884 and the other three a month later, numbered from 52 to 55, all painted in Midland Railway red. They seem to have stayed this way until the S&D adopted a blue colour scheme, which began being applied to engines during 1886.

Visually and internally there was little to choose between the Avonside and Vulcan engines, but

**Vulcan built** 0-4-4T No. 53 captured by the photographer on a gloomy, wet day at Bridgwater in 1905. This engine received a new boiler during that year and another in 1926, four years before being withdrawn from service. (Author)

**For those** of a superstitious frame of mind, engine No.53 seemed to become something of a Jonah to the men who worked on her footplate. Only bad luck of course, but in a very short period, it was involved in three accidents. On 27 February 1885, it was shunting loaded trucks at Shepton Mallet when it appears that confusion over the position of the reversing screw and double-handed arm contributed to a simple, but avoidable accident. The driver left the cab to converse with a signalman leaving the fireman on the footplate and then watched as what he had hoped was his static engine moving off pushing the wagons through the buffers and off the track. Five months later, on 13 July, the engine was again in the wars when pulling the 11.45am passenger train from Bournemouth to Bath. At Binegar Station, a temporary measure was put in place to allow fitters to undertake improvements to be made to signal box locking bars. All would have been well except that the signalman was left to operate without the usual safeguards in place and left a set of points open. The approaching train with No. 53 in charge slid over on to the adjoining track where the 2.06pm down goods pulled by 0-6-0 No. 48 and banked by No. 5 was standing. As a result of the collision that followed one passenger and the guard on the passenger train died and another sixteen people were injured. The lithograph above captures the scene as imaginatively 'witnessed' by the artist. But this wasn't the last of this engine's run of bad luck. On 3 February 1886, it was pulling thirty-three loaded and unloaded wagons plus two brake vans with the assistance of engine No.39. As it approached Binegar in fog, a goods train was allowed to enter the single track section where it met the other train head on. In the collision, both trains were damaged extensively and No. 53's fireman died. (Author)

**Engine No.** 55, the last Johnson 0-4-4T to be delivered to the S&D but not the last of this type built for the Midland Railway. This was the 2228 Class, the last of which appeared in 1900. (Author)

**A picture** of an unidentified 0-4-4T pulling into Radstock Station in the years before the Great War. The novelty of a photographer looking down on the scene is obviously of greater interest to the people on the platform than the commonplace sight of another train approaching. (Author)

there were some small differences. Generally, these reflected the constant refinement of ideas of any good design team seeking to enhance performance. Nothing is ever the finished article but only part of an evolutionary process seems an appropriate maxim in these cases and the Midland team led by Johnson did not stand still or rest on their laurels. Amongst other things, the Vulcans had steam brakes and their side tanks were higher and extended further forward allowing their water capacity to be increased from 876 to 950 gallons. The cabs were made longer with doors fitted to offer a little more protection for the crew. All in all, these modifications were fairly minor, but their overall affect was to reduce the weight by 4cwt to 43 tons 7cwt.

This process of change went on and during the lives of all thirteen 0-4-4Ts there would be a number of other changes, most notably involving the boilers. The Avonside engines were reboilered for the first time between 1889 and 1894. These new Derby designed boilers were fitted with 246 x 1⅝in tubes, instead of 219 x 1¾in tubes. At the same time, the grate area was reduced from 15sq ft. to 14¾sq ft. These changes helped raise their total heating surfaces from 1,195sq ft to 1,251sq ft, presumably with the aim of improving their steaming qualities and possibly their economy as well, but left the pressure at 140lb psi. These boilers remained in place until 1906, when a second replacement programme, lasting until 1910, was set in motion. On this occasion the tubes were reduced to 196 x 1¾in in number and diameter and were served by a smaller grate area of 14.6sq ft. These changes reduced the total heating surface to 1,074sq ft, but the boiler pressure was increased to 160lb psi. By this stage, Richard Deeley had superseded Johnson as Locomotive Superintendent and the new boilers, perhaps erroneously, came to be named after him. However, the design probably had more to do with his predecessor's work.

In between these two boiler change programmes involving the Avonsides, the Vulcans also began receiving new boilers as well. Between 1902 and 1907, they were fitted with Derby built units similar to those on the first nine locomotives, so contained the same number of tubes, plus the same heating surfaces and grate area. However, on this occasion the barrel was constructed in two sections of steel, with the dome on the forward part. In this condition ten of these engines advanced into the last decades of their lives, but the remaining three would undergo one final boiler change. In the 1920s, some Henry Fowler designed boilers of the Belpaire type were being discarded and it was decided to fit three of these to 0-4-4Ts Nos. 32, 53 and 55 during 1925/26, presumably as an experiment. These boilers were described as G5½ types because they were an amalgam of the G5 and G6s. The 4ft 11 15/16in firebox of the former was married to the 10ft 6 1/16in barrel of the latter.

Johnson had been a late arrival on the scene when it came to the Belpaire firebox, which had its genesis in the hands of Alfred Belpaire during 1864 in Belgium. Despite this, it found little use for many years, but in the early 1880s an example appeared in the USA and from that other types were gradually developed. Most notably in Britain, George Churchward, attracted by this system's free steaming qualities, developed a version of this firebox and married it to a tapered boiler as an experiment during 1902. It was fitted to engine No. 3405, a member of the Atbara 4-4-0 Class, and proved so successful during testing that within a year the idea had been woven into the design of his outstanding City Class engines.

Hamilton Ellis suggests that Johnson had been interested in the Belpaire firebox during the last years of the nineteenth century but didn't consider its application until the size of engines had increased sufficiently to justify its use. In his case, it came with the development and introduction of five Class 3 4-4-0 express engines, Nos. 2606 to 2610, in 1900/01. They proved to be sound engines and eventually their number was increased to eighty by 1905. It was a cause willingly taken up by Deeley who continued its evolution, then passed on to Henry Fowler when Works Manager at Derby before promotion to Locomotive Superintendent in 1909, then LMS's CME in 1925. It was under his guiding hand that three of the S&Ds 0-4-4Ts received their G5½ boilers and fireboxes, perhaps not the most suitable engines for this work, if one follows Johnson's thoughts on locomotive size. But the work went ahead, and the three engines returned to Highbridge. Here, in 1928, No. 32 became 52, the Vulcan engine with the same number having been withdrawn from service in May of that year. By 1932, it had been joined by the rest,

**The last** 0-4-4T to remain in service, here in its final form with the G5½ boiler/firebox combination. No. 32 became 52 in 1928 then 1230 in which state she was scrapped in 1946. Another year or so in service and the engine would have been the responsibility of British Railways. (D. Neal)

No. 32/52 remaining in service until 1946 providing a fitting memorial to a class that did much to keep the S&D going through many difficult years.

Although there were concerns amongst managers that any increase in traffic over the line might overstretch the 0-4-4Ts, they appear to have coped remarkably well with their workload for many years. And alongside them sat a few remaining locomotives from the pre-1874 days, plus the six John Fowler built 0-6-0 tender engines and nine Fox, Walker 0-6-0 tank engines. Together they didn't create an embarrassment of riches, far from it, but it was a more balanced fleet than the railway had previously enjoyed. Nevertheless, if traffic continued growing, and every indication was that it would, more motive power was needed, particularly when it came to the movement of goods. By this stage, the Midland Railway's building programme had produced 120 new 0-6-0 tender engines with many more planned. Such was the importance attached to the construction of these engines that inevitably some would be assigned to the S&D and, in 1878, six Neilson built engines arrived. A year later, they were joined by three Vulcan Foundry engines and then over the years, up to 1922, more were gradually allocated or re-allocated by successive Locomotive Superintendents. After that, the LMS took over, but in the meantime these valuable and numerous engines made their mark on the S&D.

Despite their best endeavours, the Fowler engines couldn't meet all the company's heavy main line freight needs. As a short-term

**By June** 1878, when the first of the Johnson designed, Neilson built 0-6-0 tender engines arrived, the six 'hire purchase' John Fowler engines of the same type were fully immersed in day to day activities on the line, both goods and passenger duties judging by the few photos that still exist. In this picture the first of the Fowler engines is being checked over before another turn of duty begins with steam to spare judging by the pressured steam emanating from its dome. (D. Neal)

measure, until the MR's production programme could generate sufficient numbers for its own and the S&D's needs, two older 0-6-0 engines, Nos. 351 and 353, were loaned during December 1875. These were both Kirtley designed locomotives of the 230 strong 240 Class built at Derby between 1850 and 1863 and rebuilt by Johnson a few years later. They were tried, tested and regarded as sound engines, but at this stage appeared to have been seen as inferior to the new Johnson engines, although slightly similar in appearance. Nevertheless, as a stop gap measure, they did their job and, in due course, returned to the MR when more 0-6-0s became available.

Specifications for six new engines were prepared at Derby, which were based, it has been contended, on the Midland Railway's 1102 Class 0-6-0 tank engines. Twenty-five of them were constructed by Neilson and Co of Glasgow and fifteen by Vulcan in 1874/75. Tenders for these engines were issued in May 1877, with a contract being awarded to Neilson's. By this stage, the company was a well-established engineering concern of some forty years standing, having begun under a partnership of Stewart Kerr, James Mitchell and William Neilson, with marine and stationary engines as their raison d'être. But this soon changed with the rapid expansion of Britain's railways. In 1843, they began producing 0-4-0 steam locomotives, ending the marine side of business twelve years later, such was their success in this new market. These dates

**0-6-0 No.** 323 it seems was one of 230 Kirtley 240 Class engines all built at Derby between 1850 and 1863. In his 1953 book *The Midland Railway*, Hamilton Ellis suggests that these engines belonged to the later 237 strong 480 Class, assigning their construction to Derby and a number of contractors. Later research seems to suggest that they are probably 240 Class locomotives though this isn't entirely clear to me. Either way, two Kirtley 0-6-0s, Nos. 351 and 353, were loaned to the S&D to help them over a difficult period until new Johnson engines could be supplied. It seems that they had 5ft 2in. coupled wheels, 16in by 24in cylinders, increased to 16½in and then, in some cases to 17in, by Johnson. They were fitted with relatively shallow plate frames with the frame tops and platforms curved over the axleboxes. Although remaining with the S&D for only a comparatively short period, they appear to have worked effectively over the line. (D. Neal)

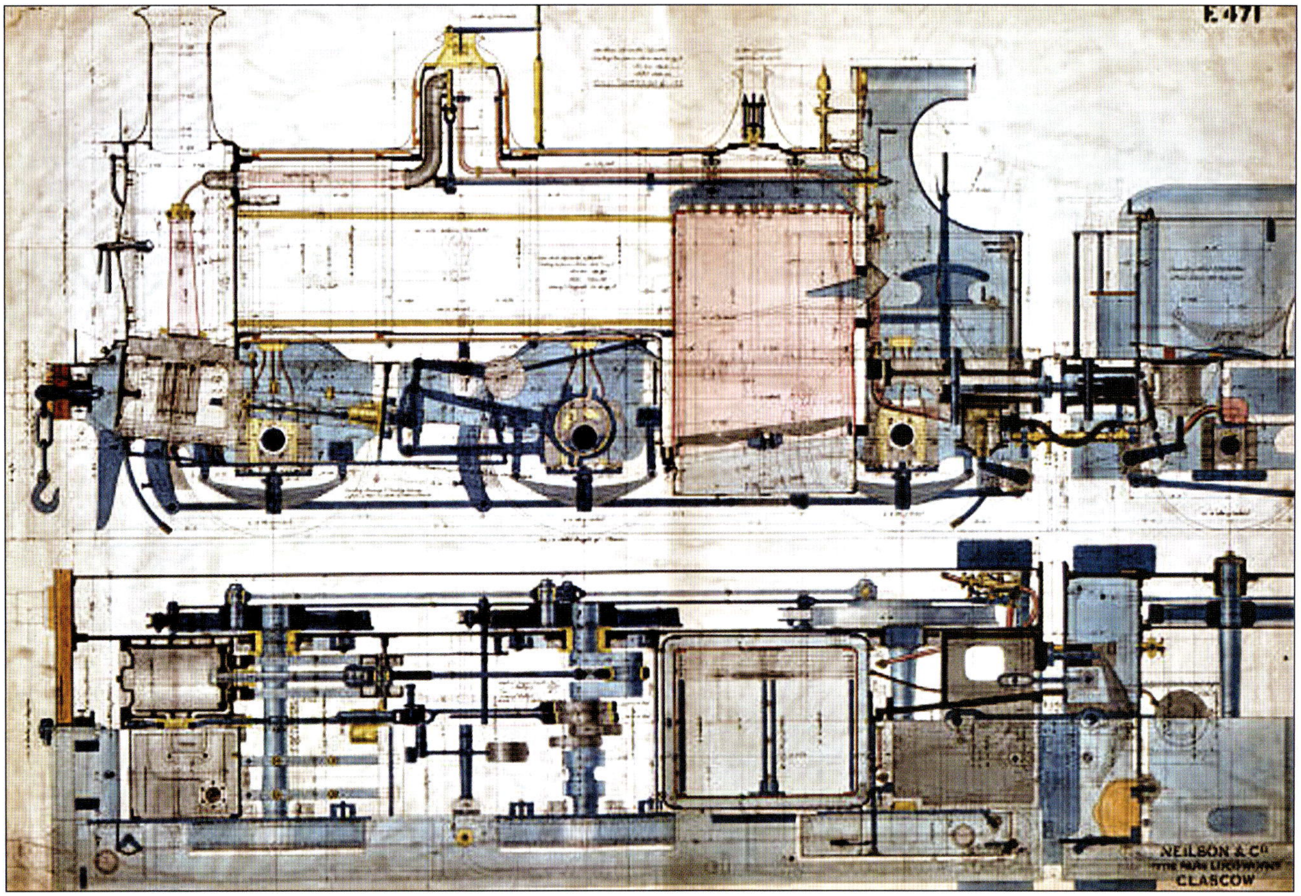

**One of** Neilson and Co's colourised drawings produced to accompany the order they received in 1878 for six 0-6-0 goods engines. The programme didn't run smoothly, though, and there were potential production delays that meant that Neilson's sought a delay. Under the threat of cancellation and possible legal action the company rallied round and the engines were all delivered in June/July 1878. (D. Neal)

coincided with changes of name to Mitchell and Neilson then simply Neilson and Co, with William Neilson's son Walter eventually taking charge. It was he who then expanded the business, quickly adopting the business mantra 'Scottish locomotives for Scottish railway companies'. By 1876, Walter had formed a new partnership with his brother George and was constructing machine tools in the nearby Albert Works, which left James Reid, formerly Works Manager, to manage the company and then become sole proprietor when Walter withdrew from the business in 1878.

The contract for these engines neatly fell in the middle of these changes, which may explain a small hiccup in their construction and delivery. Having accepted Neilson's bid of £2,275 for each locomotive, it was expected that delivery would be achieved by mid-1878. However, in the face of problems with the construction programme, Reid sought an unspecified, but probably short extension during March to allow the tenders and boilers to be built by sub-contractors.

A dispute followed, during which cancellation of the order became a real possibility, but with no other company able to produce these locomotives in the right timescale and for less than £2,450, their hands were tied. There would also have been the question of a costly court action and possibly compensation to consider. So Neilson's pressed on and all six were delivered by July, no more than a month or two late. However, when it came to order more 0-6-0s, very noticeably the work went to the Vulcan Foundry or remained in the precincts of Derby.

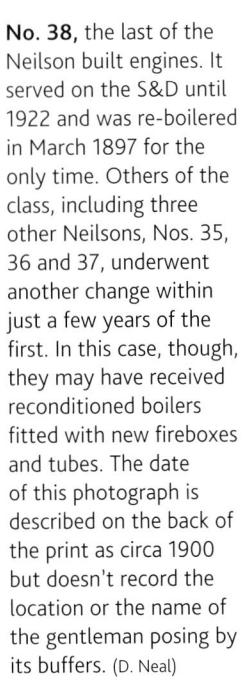

**No. 38,** the last of the Neilson built engines. It served on the S&D until 1922 and was re-boilered in March 1897 for the only time. Others of the class, including three other Neilsons, Nos. 35, 36 and 37, underwent another change within just a few years of the first. In this case, though, they may have received reconditioned boilers fitted with new fireboxes and tubes. The date of this photograph is described on the back of the print as circa 1900 but doesn't record the location or the name of the gentleman posing by its buffers. (D. Neal)

**This view** of No. 37 (built as No. 60 but re-numbered when the original No. 37 was withdrawn from service in 1922) was taken at Evercreech Junction in 1929 three years before withdrawal. It captures the balanced look of these Johnson designed locomotives. By this stage, the new No. 37 had been reboilered twice, the first time in 1908 and then in 1921 when it was fitted with a G5 Belpaire boiler and new frames. Hamilton Ellis described Johnson's engines as 'a work of art and made like a watch. Mechanical judgement was their virtue'. These 0-6-0s seem to bear out his view – form and function in harmony. In terms of dimensions, they had 17in x 24in cylinders, 4ft 6in coupled wheels, a 5ft long firebox with a 15ft grate area, a 10ft long boiler with a diameter of 4ft 2in containing 215 x 1¾ in tubes, and a total heating surface of 1,124sq ft producing a working pressure of 140lb. The locomotive and tender had a weight in working order of 62 tons 11cwt. To all intents and purposes, they were similar to other 0-6-0s being built by the Midland Railway for its own use but were smaller in certain respects – the wheelbase was 1ft 6in shorter, the boiler barrel less by 6in and the firebox 11in less.

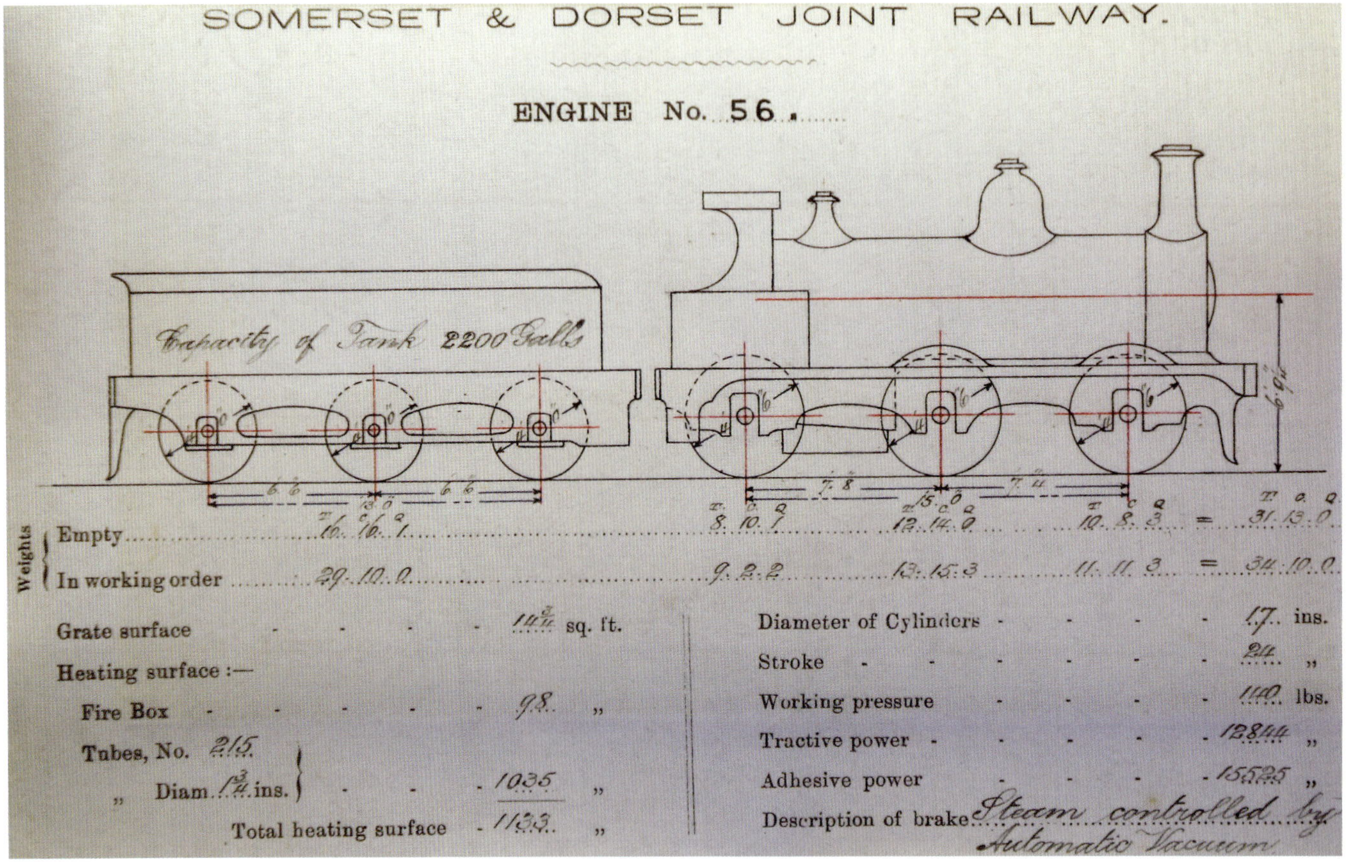

**A typical** example of a Vulcan built Johnson designed 0-6-0, in this case No. 56 which was delivered to the S&D in June 1890 with the works number VF 1264. It was the first of six locomotives (56 to 61) to be fitted with vacuum ejectors so that it could be used to pull passenger trains when required. This modification was quickly extended to all other members of the class. No. 56 lasted in service until October 1928, by which time eleven of the larger more powerful Fowler 2-8-0s were in service. (Author)

It wasn't until 1902 that Neilson and Co again supplied any 0-6-0 engines to the S&D, but by then their name had changed once more to become Neilson, Reid and Co.

Once these engines were in service, it wasn't long before the S&D expressed a wish to acquire more 0-6-0s and tenders were invited from a number of companies. Vulcan of Manchester submitted a bid in March 1879 which rated each engine at £2,090. This being much lower in price than Neilson's valuation of £2,275 for the first six locomotives suggests that Vulcan were determined to win the competition. Delivery was scheduled for later that year, but it seems that a bottleneck in their production schedules made this difficult to achieve. They tried to negotiate a six-month delay, sweetened, or so they hoped, with a £45 discount on each engine, but were rebuffed by the customer and threatened with cancellation. This focussed their minds on the task at hand and the first three were delivered in December and the final three a month later. Despite this early hitch, Vulcans continued to win contracts for more of the class and by 1890 the total delivered had risen to twenty-two, with various groups of numbers allocated – the 1879/80 batch were 39 to 44, the four in 1881 became Nos. 25 to 28, six in 1884 were Nos. 46 to 51 and the final six in 1890 were given the Nos. 56 to 61. However, along the way, and due to the Glasgow origins of the first six, the crew who worked these engines nicknamed them 'Scotties'. It was a soubriquet that attached itself to this entire group of locomotives, whether Caledonian or Mancunian in origin.

These new locomotives are said to have been based on the 1377 Class 0-6-0 tank engines which were a development of the earlier 1102 Class. In this guise, Robin Atthill concluded that they were the only fully standard Johnson designed Midland engines allocated to the S&D. The 1377s first appeared in 1878 and over the next fourteen years their total number rose to 185, the last group of twenty being built by Vulcan in 1892. They may well have been based on these tank engines, and they do bear a passing resemblance to them, but some of the basic dimensions are different, as they are to the Neilson 0-6-0s. For example, the 1377s had coupled wheels of 4ft 7in, 17in x 24in cylinders and a boiler working pressure, available records suggest, of 175lb. The Vulcan engines' comparable dimensions were 5ft 2½in coupled wheels, 18in x 26in cylinders and a pressure of 140lb. By comparison, the Neilson engines had 4ft 6in wheels, 17in x 24in cylinders, with a boiler producing 140lb of pressure. From this, it is probably safe to say that they all came from the same design team and so were all part of one evolutionary process, but beyond that it seems to be more speculation than certainty. In any event, the Neilsons, no matter where their antecedents lay, were designed and built specifically for the S&D whilst the Vulcan engines were part of a more numerous MR standard design.

In service, all the 0-6-0 goods engines, whether tank or tender, were for safety reasons set load limits over the more difficult sections of line. Bath to Evercreech was the most obvious of these with its steep inclines. All versions of the 0-6-0s, whether Fowler, Neilson or Vulcan built, were limited over this section to twenty-six goods, twenty-two coal or thirty-seven empty trucks, plus goods vans when banked by another engine. These totals were reduced to fifteen, twelve or twenty-three unassisted. On easier sections south of Evercreech to Wimborne, numbers were increased to thirty-five, twenty-seven and forty-five. This was far better than other types available until the larger, more powerful 2-8-0s turned up during the Great War, but marginally less than the Fox, Walker 0-6-0STs on either section. However, their range was limited by the amount of coal and water they could carry and so, post-1890s, as the tender engines became more numerous, their main line duties with such heavy loads diminished. However, the one exception to this was the Fox, Walker ST engine No.1 which in January 1888 was turned out as a tender engine, so giving it the range of the other 0-6-0s and presumably the pulling power of the other tank engines. It remained in this state for twenty years until restored to ST status, in which condition it remained until withdrawn during November 1930.

Although there is some colloquial evidence that the men who worked the 0-6-0s had a preference for the Fowler built engines, finding them stronger and more reliable than those that came later, the Neilson and Vulcan locomotives soon displaced them on many of the most demanding services. It was felt that they steamed more freely, and the brakes were more effective on the challenging descents on

**The first** Fox, Walker 0-6-0ST as rebuilt in 1888 as an 0-6-0 tender engine, presumably because the company had sufficient tank engines at the time, even with twenty-two Neilson and Vulcan 0-6-0s on the S & D's books. In 1908, this engine was returned to its original saddle tank form and is pictured here in LMS markings and numbered 1500 in which state it was scrapped in 1930. (D. Neal/Author)

**Derby in** 1890 with various Kirtley and Johnson locomotives on display in No. 4 shed. Of particular note is S&D 0-6-0 No. 60, recently built at the Vulcan Foundry and at Derby, whilst running in before transfer to Highbridge to begin work. Of particular note is the uncluttered layout of this shed and the apparent high standard of cleanliness. (D. Neal)

the line. Economy may also have been an issue. Nevertheless, the Fowlers continued to serve the S&D, without major reconstruction until the late 1920s, though were reboiled twice in the process. In so doing, they were deemed to be a successful design and justified the decision to acquire them in 1874. The same can be said of the Vulcan built engines, which continued to prove their worth into the 1930s. This probably encouraged the company to acquire more and another five built at Derby arrived

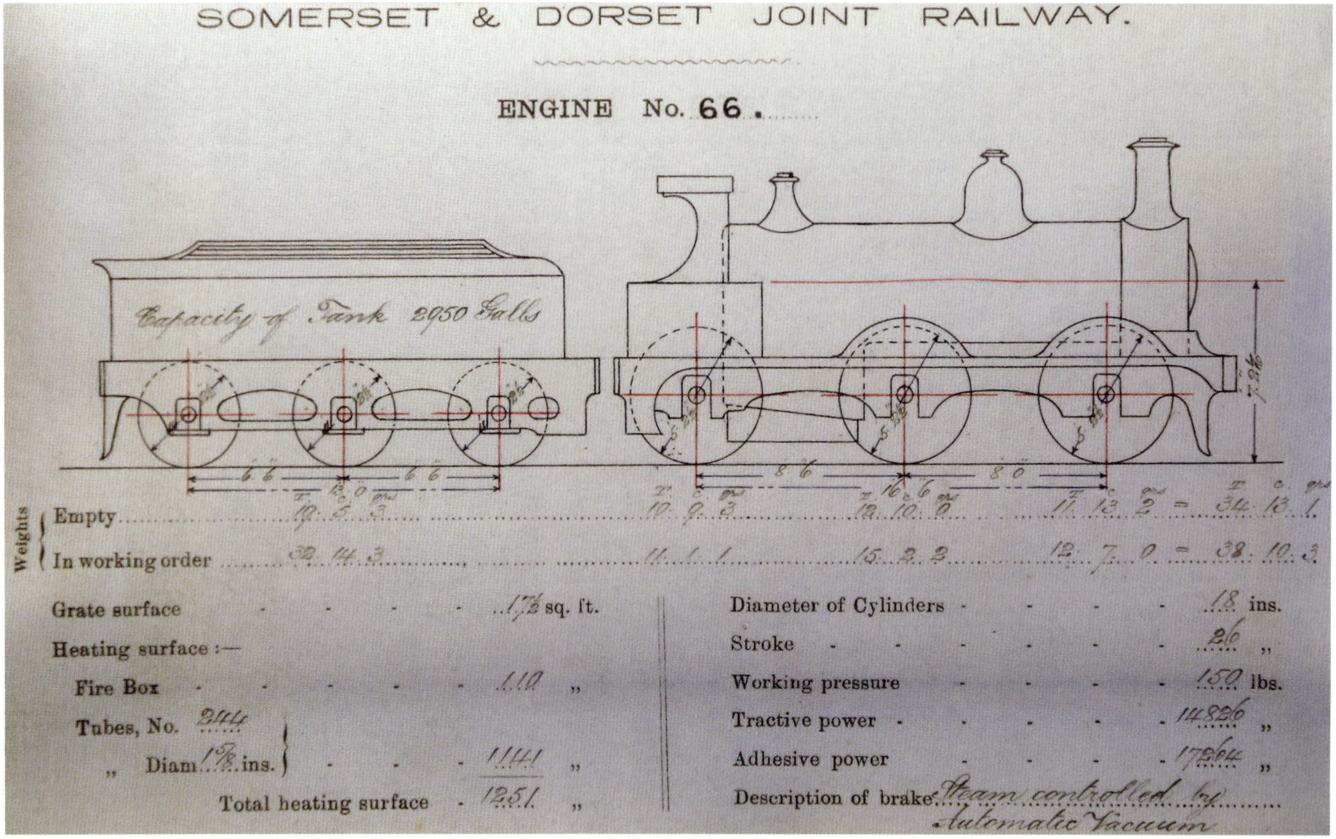

**In 1896,** an additional five 0-6-0s were built for the S&D at Derby (Nos. 62 to 66). The specification called for locomotives with more power than those already in service, which, during the winter months were struggling with heavier loads. When bidding for this work, the MR were in competition with the LSWR. With a price of £2,245, which included the manufacturer's profit, they won the contract. Essentially, they were bigger versions of the earlier 0-6-0s. So, for example, they had 18in x 26in cylinders, 5ft 2½in diameter coupled wheels, a boiler six inches longer containing 244 x 1⅝in tubes, a grate area of 17½sq ft, a 16ft 6in. wheelbase and could produce a working pressure of 150lb psi. They quickly proved themselves superior to the other engines in service and garnered the nickname 'Bulldogs' such was their reliability and capacity for hard work. The diagram above displays their dimensions and the photo their looks. (Author)

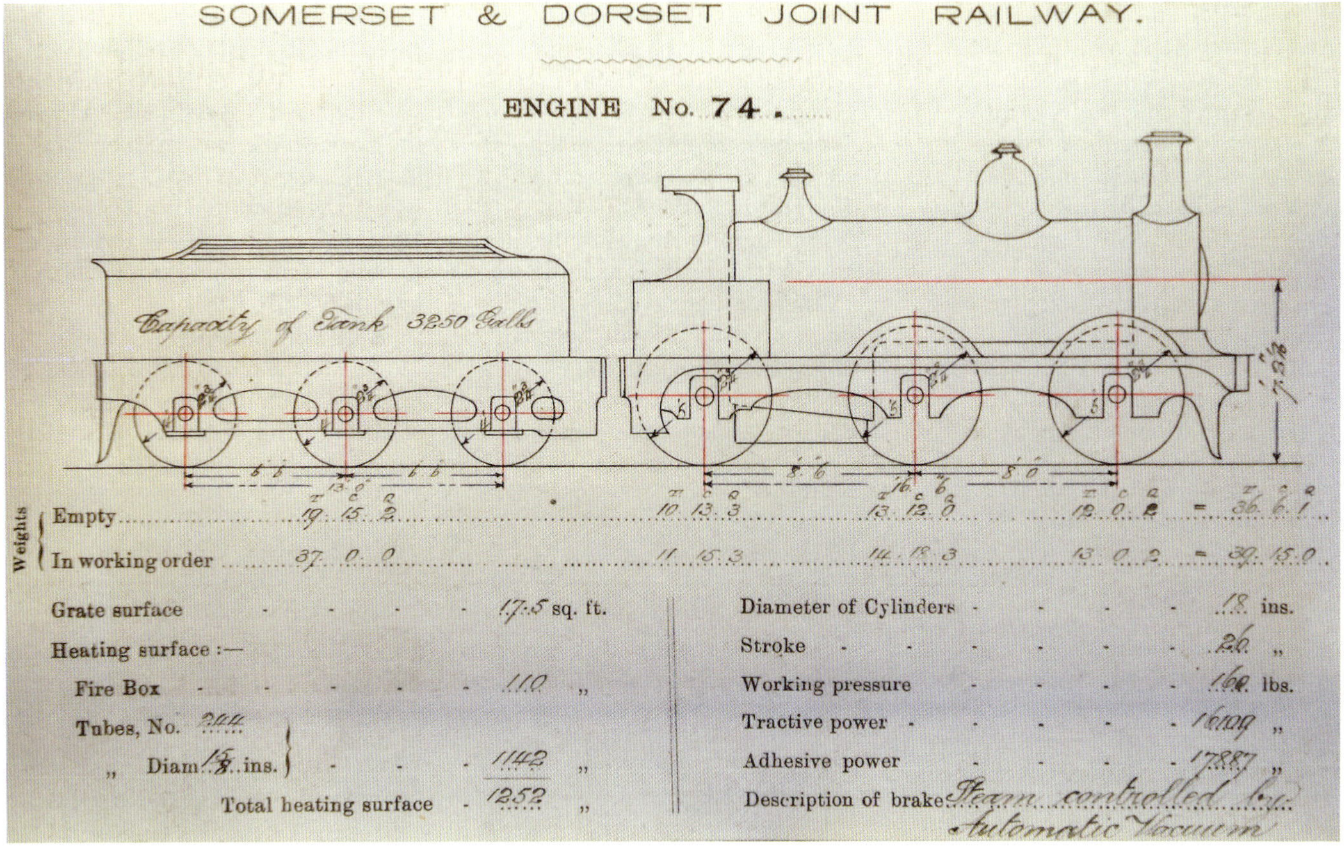

## SOMERSET & DORSET JOINT RAILWAY.
### ENGINE No. 74.

*[Engineering drawing with handwritten specifications:]*

Capacity of Tank 3,250 Galls

| Weights | | | | | | |
|---|---|---|---|---|---|---|
| Empty | 19.15.2 | | 10.13.3 | 13.12.0 | 12.0.2 | = 36.6.1 |
| In working order | 37.0.0 | | 11.15.3 | 14.19.3 | 13.0.2 | = 39.15.0 |

| | | | | |
|---|---|---|---|---|
| Grate surface | 17.5 sq. ft. | Diameter of Cylinders | 18 ins. |
| Heating surface :— | | Stroke | 26 " |
| Fire Box | 110 " | Working pressure | 160 lbs. |
| Tubes, No. 244 | | Tractive power | 16,109 " |
| " Diam 1⅝ ins. | 1142 " | Adhesive power | 17,887 " |
| Total heating surface | 1252 " | Description of brake | Steam controlled by Automatic Vacuum |

**In 1902,** Neilson, now under the name Neilson, Reid and Co, were entrusted with another order for 0-6-0 tender engines, twenty-four years after producing the first six in 1878. As the new century dawned, goods traffic had increased sufficiently to encourage the S&D to authorise the purchase of more engines. At the same time, Neilson's were in the middle of an MR contract to build sixty 0-6-0s at a cost of £3,185 per engine. To meet the S&D's requirement, Johnson simply diverted five of these engines to Highbridge. They were delivered in September 1902, painted in Midland red, having been allocated the numbers 72 to 76. There were a number of differences between them and the earlier engines, most noticeably a larger tender able to carry 3,250 gallons of water, deeper frames, a working pressure of 160lb and 5ft 2¾in diameter wheels. With a combined loco plus tender weight of 76 tons 15 cwt, they were more than 5 tons heavier than engines 62 to 66. These larger 0-6-0s, from Derby and Neilson, had a long and active service on the line with some lasting into the early 1960s. (Author)

**The third** of the Neilson, Reid 0-6-0s delivered to Highbridge in 1902. No. 74 remained in service until October 1952. However, it was No. 72 that became the longest serving member of class, only being withdrawn in August 1962 having been reboilered in October 1925 when fitted with a G7 Belpaire combination. At the same time the locomotive received new frames. Engines 62 to 66 and 73, 74, 75 and 76 were similarly treated between 1920 and 1925. (Author)

in 1896, followed by a further five from Neilson, Reid and Co in 1902. There would be a final group of five built by Armstrong Whitworths and Co of Scotswood, at their works to the west of Newcastle-on-Tyne in 1922, but more about this last group of engines in the next chapter.

The 0-6-0s' safety record, considering their large number, was fairly good. They were involved in a number of accidents, but all of these seem to have been the result of human error. There is no evidence to suggest that their design was a contributory factor, although an accident at 9.30 on Saturday 2 November 1889, involving Neilson built engine No.36, led to all tender locos being banned from shunting duties. But this accident had one other significant consequence affecting, as it did, the S&D's senior management team.

Engine No. 36 was moving coal trucks in the sidings at Highbridge by propelling them in small groups. Familiarity often breeds contempt or at least carelessness and, in this case, the Resident Locomotive Superintendent, William French, who replaced Fisher in 1883, crossed the track on the way from home to his office between two clusters of wagons without looking properly. He didn't see the engine and trucks moving towards him and walked into their path, being crushed to death between the buffers of two loaded wagons as they came together. Although in post at Highbridge for six years, his impact

**A good** example of how the larger 0-6-0 engines looked having been fitted with a G7 boiler and Belpaire firebox, in this case Derby built No. 64 which was turned out in this state in February 1921. According to the information attached to the negative, this photo was taken in April 1921 at Midford with a train to Bath. (D. Neal)

| Name in full | Whitaker, Alfred | | Date of Birth | July 22nd 1846 | |
|---|---|---|---|---|---|

| Station | Date | Appointments and Advances | Salary or Wages £ s. | |
|---|---|---|---|---|
| Derby | Oct 15 60 | Employed as Clerk | · | 7 |
| „ | Oct - 61 | Advanced to | · | 10 |
| „ | Dec - 62 | „ | · | 14 |
| „ | Jan - 64 | „ | · | 18 |
| „ | Sept 26 64 | To Fitting Shops | | |
| „ | Jan 22 68 | Advanced from 21/- to | 1 | 7 |
| Lancaster | July 29 69 | Locomotive Foreman | 100 | · |
| „ | Feb 2 70 | Advanced to | 125 | · |
| „ | Jan 1 71 | „ | 140 | · |
| Bradford | Apl 1 71 | Locomotive Foreman | 150 | · |
| „ | July 1 72 | Advanced to | 160 | · |
| Derby | Nov 19 72 | Appointed Assistant District Inspector | 170 | · |
| Leeds | June 17 73 | Appointed Locomotive Foreman | 250 | · |
| Carlisle | July 75 | „ | 250 | · |
| Leeds | Jan 1 81 | „ | 270 | · |
| „ | July 1 87 | Advanced to | 290 | · |
| Highbridge | Nov 14 89 | Apptd Resident Loco Supt of the S&D Joint Ry vice French (Ja 2316) | 375 | · |
| „ | June 1 92 | Advanced to (SD 5065) | 425 | · |
| „ | July 1 96 | „ „ (13125) | 475 | · |
| „ | Feb 1 1900 | „ „ (S.D Min: 2425) | 500 | · |

To be retained in the service. Minute 5960
(S&D) 22/1/07
660/37

**In 1889,** following William French's sudden death, Alfred Whitaker was appointed to be Resident Locomotive Superintendent at Highbridge. He brought a new dynamism to the work of his department, something that seems to have been lacking since the sudden death of Benjamin Fisher in 1883. By the time of Whitaker's appointment, he had acquired considerable experience of loco management as his employment record reveals. His son, also called Alfred, would also play an active part in the work of the S&D, having moved from Derby to the S&D in 1907. During the Great War, he became acting Locomotive Superintendent, reverting to assistant in 1919. (Author)

on the procurement programme was considered to be a fairly slight one. Six Vulcan built 0-6-0s appeared, plus four 0-4-4 tank engines and a single 0-4-2ST, No. 25A, for shunting duties at Radstock but that was all. In addition, there was a view forming that standards had slipped during his period of tenure. There will inevitably be many reasons for such apparent inactivity on the locomotive front – the fleet being sufficient for the company's needs probably being the most obvious. However, it has been suggested that he was a pressed man when taking up the appointment, didn't wish to live in Somerset and this unhappiness may have been reflected in his performance. But in the business world, life moves swiftly on and the appointment of Alfred Whitaker, then based at Leeds, was suggested and soon ratified. He quickly set to work and soon established that all was not working as effectively as it should be and moved swiftly to implement change. This included the introduction of better working practices across his department, the development of the existing workshop facilities at Highbridge and the acquisition of more Derby constructed locomotives, to sit alongside those engines built by

**Although not** trained as a designer Alfred Whitaker harboured design ambitions which he was able to exploit at Highbridge in developing the facilities there. One example of his work could be found in the Erecting Shop. The overhead travelling cranes had insufficient capacity to lift the heavier engines; to avoid the cost of laying new foundations and fitting new cranes he designed a highly successful portable incline traverser, as seen here forward of Vulcan Foundry built engine No. 59. (Author)

contractors. Spreading the risk perhaps?

In pursuing these various programmes, it undoubtedly helped that he had developed a close working relationship with Johnson and his team at Derby. This probably meant that he was better placed to enlist their support where necessary for the work he thought essential. However, change doesn't happen overnight and inevitably follows a period of study and analysis to determine where weaknesses might lie. But, within a year, he had begun to enhance technical training to a level comparable with that of the Midland Railway, which quickly led to an improvement in working standards and output. The workshops at Highbridge then became the target of his modernisation programme. Some of the buildings were extended and then equipped with machinery from Derby, albeit second hand, though much better than they had become used to. All in all, it was a complete makeover of an organisation that was functioning but had not always adopted best practice, modern techniques or truly effective ways of working.

Whitaker was more than a production engineer; he also had something of the inventor about him. Though unable to develop these skills with the MR to any great extent when a Locomotive Foreman, signs of inherent design skills emerged when working at Highbridge. One example of this was a portable incline locomotive traverser he produced for the Erecting Shop. He found that the building's structure could not support an overhead crane of sufficient capacity so set about designing an alternative solution. To this could be added his highly-successful token exchange apparatus, his water tank depth gauge, both of which found use across the industry, and a spark arrester gear, which didn't.

During his very active years at Highbridge, a number of new engines arrived. There was, of course, the continuing build-up of 0-6-0 tender engines and two additional 0-4-0STs for Radstock. But it was probably the appearance of the first 4-4-0 passenger engines to work the line that had the biggest impact. By 1889, Johnson's development of engines with this wheel configuration was well advanced and ninety were already in service with the MR by then. It was a very active development programme and by 1899, a further 185 would be added to their inventory. With so many in service, and apparently performing well, it wasn't surprising that some found their way to the S&D where increases in passenger traffic and tighter schedules called for something more advanced than they had already.

Between 1891 and 1908, thirteen of these engines would make their appearance on the S&D, ten by 1903 when Johnson's term of office came to an end. Considering the size of the locomotive fleet, and the compactness of the line, this comparatively small number would become a dominating presence in the years up to the Second World War such was their impact. Over the years, the design of these engines would continue to evolve, so reflecting the evolutionary nature of the design team's work at Derby.

Having only been in post for a short while, Whitaker began to express his doubts about the

'**They also** serve' is an appropriate description of the minor classes of engines that do their work in quiet corners of a railway network away from the cut and thrust of main line duties. The S&D had its own small group of such engines, all of which were acquired to support the mines and sawmills around Radstock. The first of these, which is pictured above, was an 0-4-0ST built by Slaughter, Gruning & Co in the 1850s for service in chalk quarries in Essex. It arrived on the S&D after working for various other business concerns until purchased by Fisher for £385 during 1882 in a fairly dilapidated condition. After modification, and being named *Bristol*, it served at Radstock as No. 45 then 45A until 1895 when it 'was laid aside' as being no further use to the company and replaced by two new 0-4-0STs. (D. Neal)

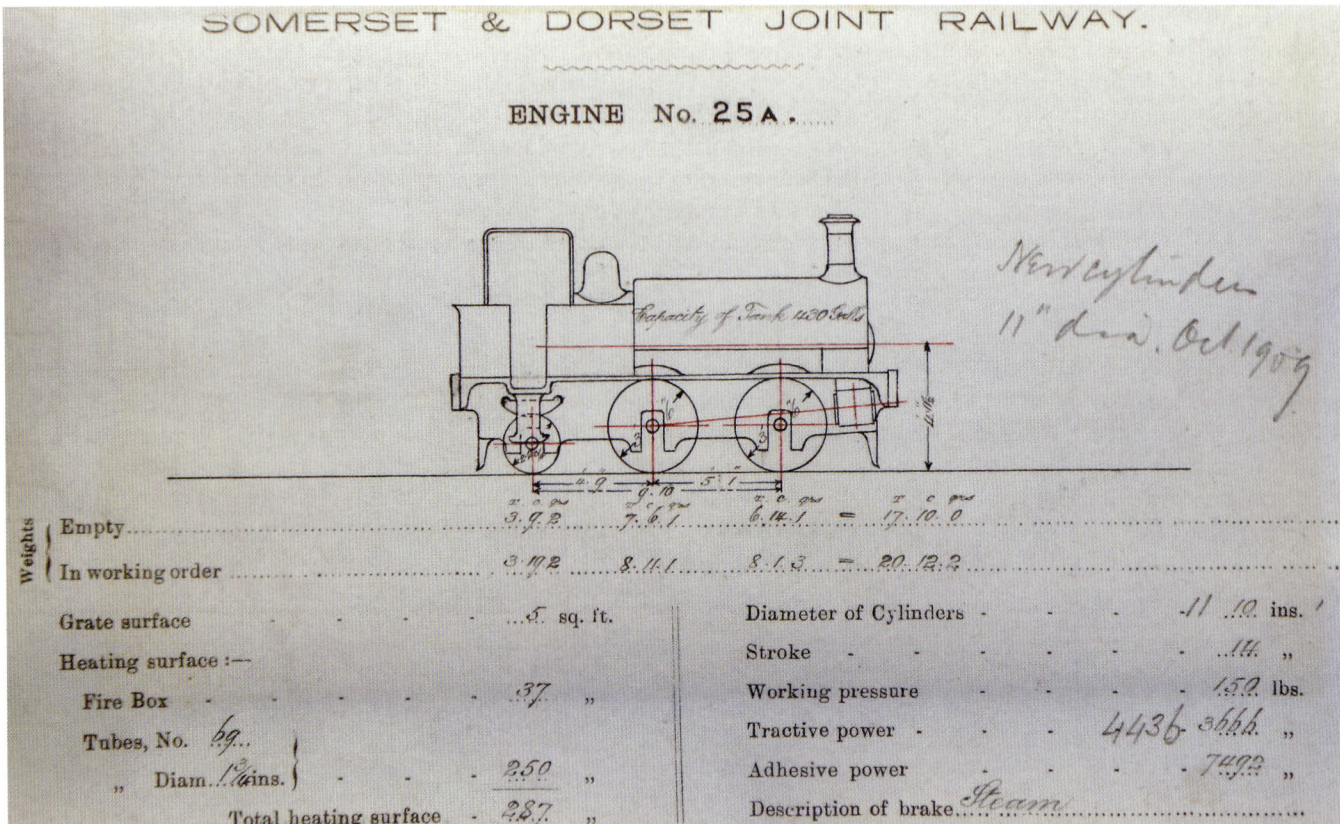

**During 1885,** No.45 acquired a new partner in the form of an 0-4-2 outside cylinder saddle tank engine given the number 25A. Most unusually for the S&D, it was constructed in the works at Highbridge at a cost of £1,095. It is illustrated above as it appeared in 1900 after a rebuild in 1897, which saw the saddle tank extended to the front of the smokebox and the cab modified. (Author)

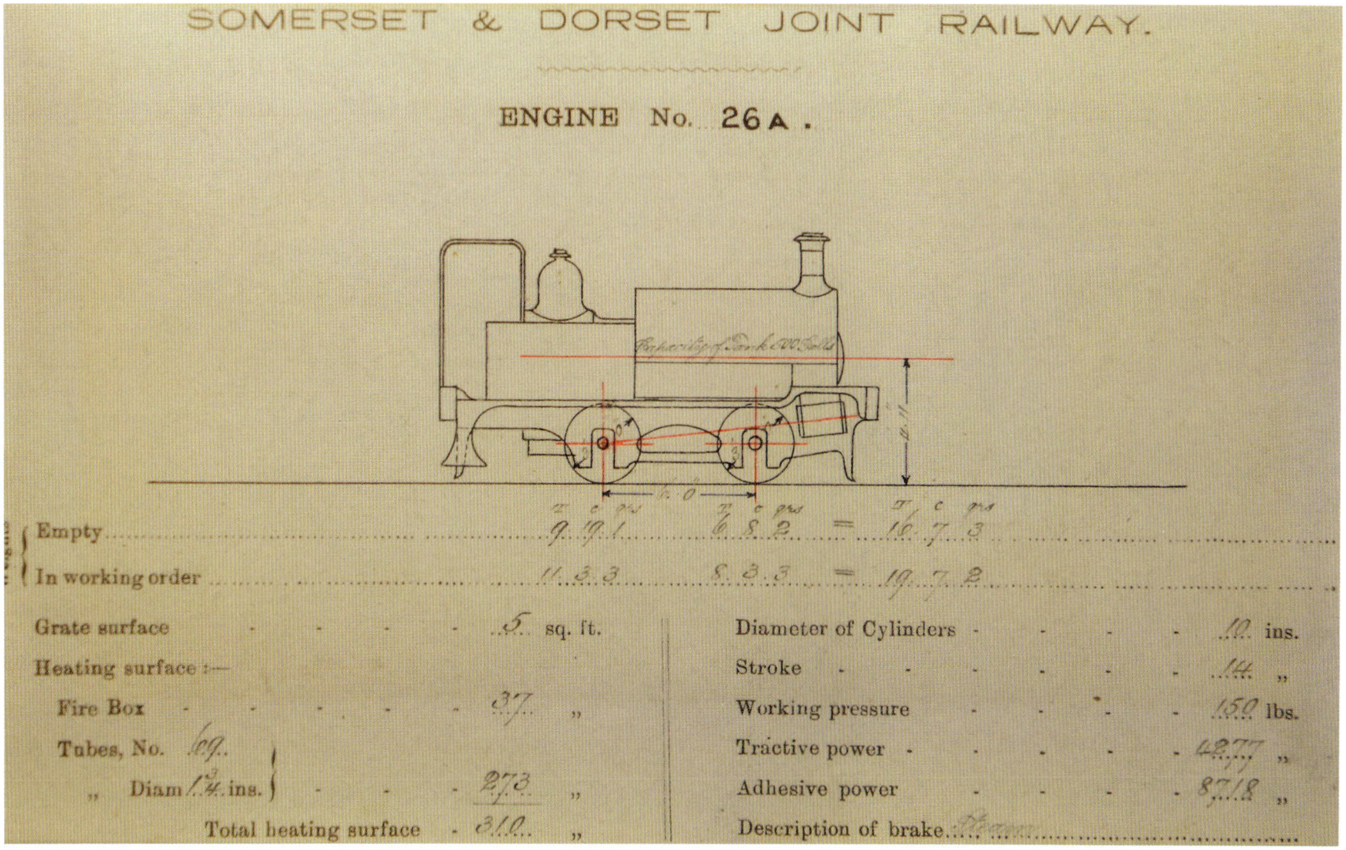

## SOMERSET & DORSET JOINT RAILWAY.

### ENGINE No. 26A.

| | | |
|---|---|---|
| Empty | 9.19.1 | 6.8.2 = 16.7.3 |
| In working order | 11.3.3 | 8.3.3 = 19.7.2 |

| | | | | |
|---|---|---|---|---|
| Grate surface | 5 sq. ft. | Diameter of Cylinders | 10 ins. |
| Heating surface :— | | Stroke | 14 " |
| Fire Box | 37 " | Working pressure | 150 lbs. |
| Tubes, No. 69 | | Tractive power | 4877 " |
| " Diam 1¾ ins. | 273 " | Adhesive power | 8718 " |
| Total heating surface | 310 " | Description of brake | Steam |

**The third** and fourth saddle tanks were 0-4-0s and were delivered to Radstock in April (No. 45A) and October (No. 26A) 1895 respectively from Highbridge. They are pictured above as they appeared in 1905. The third photograph shows 26A at work at the Radstock sawmills with, according to the text accompanying the negative, Driver Pitman in charge. All these engines were given the S&D's full passenger blue livery and were kept in a highly polished condition by their crew so attracting the nickname 'Dazzlers'. 25A and 45A lasted in service until February and August 1929, with 26A going in December the following year. It is interesting to note that all the Radstock shunting engines were chosen because of their low profiles which allowed them to enter the colliery sidings which were accessed by passing under Tynings Bridge, nicknamed Marble Arch, which had a clearance of only 10ft 10in from rail level to the roof. (Author)

**Whitaker's ascendency** presaged the arrival of Johnson's 4-4-0 tender engines for passenger duties. The sixth of these, No. 68, to reach the S&D is captured here in 1906. This locomotive was built at Derby in 1896 and survived in service until the mid-1920s. This picture captures the elegance of design at this time, the quality and finish of the livery and the standard of dress expected of the footplate crew. (Author)

use of the 0-4-4 tank engines for main line passenger duties. They had performed well since their introduction in 1877, and as late as 1884/85 French had been happy to acquire four more. But flange wear on the leading tyres was excessive and they needed frequent maintenance to keep them operating effectively and in a safe condition. If the tyres had been of higher tensile strength, the maintenance task could have been less, but there was little prospect of them being modified in this way. In any case, Whitaker was also concerned that they lacked the capacity to meet rising levels of passenger traffic, particularly during the summer months. As a result, in December 1889 he requested four more engines and specified that they be 4-4-0s, a request that Johnson was happy to approve. They would be built at Derby and cost £1,950 each and were scheduled for delivery in mid-1890, though this slipped to May the following year due to pressure of other work.

These engines, although proving their value in service, did not immediately lead to more of the type being requested. It would take another six years before their number swelled to eight, with Nos. 67 and 68 arriving in 1896 and 14 and 45 a year later from Derby at a cost of £2,230 each. With eight available, it might be thought that the ever-increasing Bath to Bournemouth passenger traffic would have been re-timed, but it remained unchanged from its 1880 set schedule. For example, this allowed 1 hour 45 minutes from Bath to Blandford, but by the turn of the century single line sections of the track between Bath and Templecombe had been doubled up making faster running possible.

By then, the fastest 4-4-0 service ran between Bath and Bournemouth in 2hr 10min, not as big an improvement as Whitaker wanted, but better, nonetheless. This was largely due to the increased loads being pulled and the arrival of much heavier, electric-lit MR coaching stock. This put an added strain on these engines, many of which needed the assistance of a second engine so they could cope with the larger loads. It was also found that the 4-4-0s' coal and water consumption was greater than expected. When compared to similar engines working on the Midland main line, the coal burnt was on average 30 per cent higher. And to this could be added higher maintenance costs due to excessive wear over comparable mileages.

Ever aware of the problems this created, in 1900, Whitaker sought to acquire more 4-4-0s, but had his requests turned down, presumably for financial reasons. At the same time, he pushed forward his case for bigger, more powerful engines that could cope with the line's gradients better without the added expense of a second engine. However, funds were restricted and, although agreeing with his summary, the management committee couldn't acquiesce to his request initially. However, with costs ever increasing and delays in schedules becoming an ever-greater embarrassment, approval for three new engines at a cost of £3,500 each was finally granted in August 1902, with delivery set for April the following year. But here Whitaker would face another frustrating delay. Derby's workshops were unable to comply, due to pressure of other work, and the best they could

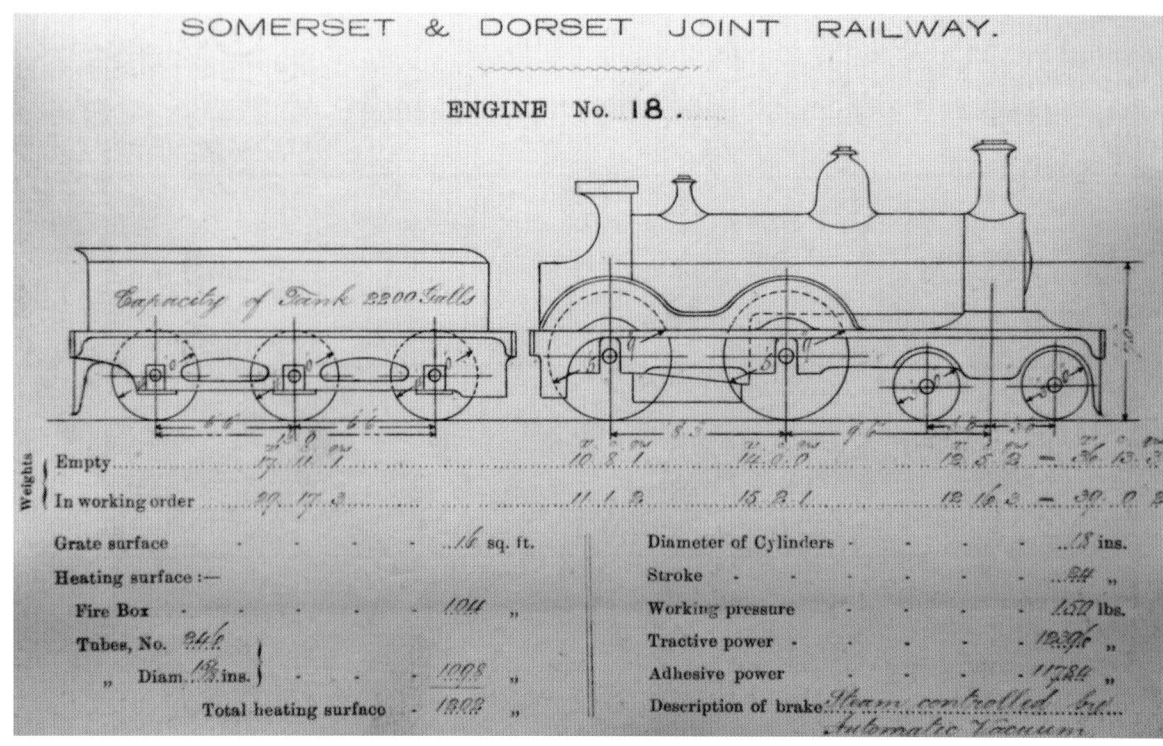

**The first** four Johnson 4-4-0 tender engines for the S&D were delivered in May 1891 and given the numbers 15 to 18. They were seen as smaller versions of the 4-4-0s employed on the MR, which themselves had gradually been growing in size. For example, by the early 1890s, their 4-4-0s had longer wheelbases, larger cylinders and coupled wheels (their size did vary but fifteen built in 1893 had 6ft 6in wheels by comparison to the S&D's engines which were 5ft 9in). In addition, their fireboxes were longer and the grate areas larger and they had working pressures of 160lb. The diagram above was prepared for the locomotive department and copies were given to workshop staff and footplate crew. The photo of No. 18 captures the balanced design of these engines but also the open working conditions for the crew. (Author)

**4-4-0 No.** 14 which arrived at Highbridge in February 1897. This engine, plus 45, 67 and 68, which were the second batch of engines to be built at Derby for the S&D, were virtually the same as the first four locomotives that arrived in 1891. However, there was a small difference in the sight-feed lubricator pipes and the way they were positioned, reflecting MR practice current at the time they were built. By 1896/97, the pipes had been moved from beneath the cladding plates to run along the left handrail. (Author)

**Engine No.** 68 as it appeared shortly after entering service in January 1896. It stayed in this condition until May 1908 when it received an 'H' round topped boiler. This engine remained in service until November 1921. (Author)

do was to give the vague estimate of 'as soon as possible'. Faced with this impasse and wishing to achieve long-awaited improvements in service, alternatives were sought, including switching production to another company or a contractor. With the LSWR as a partner in the joint railway, it was hoped that these three engines might be manufactured in their Nine Elms workshops. But the same problem of insufficient capacity existed here as well and, with no contractor of any note being able to help, the S&D were thrown back on the mercy of the Midland Railway. In due course, by adding a measure of priority, they produced the three engines in November 1903, which were given the numbers 69, 70 and 71. Whitaker wasted little time in pressing them into service on the heaviest express trains, where their performance quickly confirmed his strongly held belief that bigger engines were badly needed.

From the limited amount of information available, it seems that the drivers were happy with these three engines. They were found to steam well and were capable of pulling heavier loads than their smaller sisters, but this wasn't as great as hoped and banking on the steeper gradients was still necessary. However, the firemen were less impressed; the larger boilers and grates were found to eat coal at a far greater rate than the smaller 4-4-0s. This was confirmed later when statistics revealed that engines 69 to 71 consumed, on average, 48.9lb of coal per mile, whilst oil consumption was also greater. There were also concerns amongst the S&D team, including Whitaker,

that the coupled wheels were too large and may have benefitted from being fitted with the 5ft 9in variety attached to the earlier engines. Why this should be is not entirely clear, the extra 3in probably being insufficient to make a significant difference. However, there were

other factors in play here that might have affected performance – wear on the tyres and the occasional requirement to machine them, the skill of individual drivers and so on – for any real conclusions to be drawn. But it seems that once the size of the

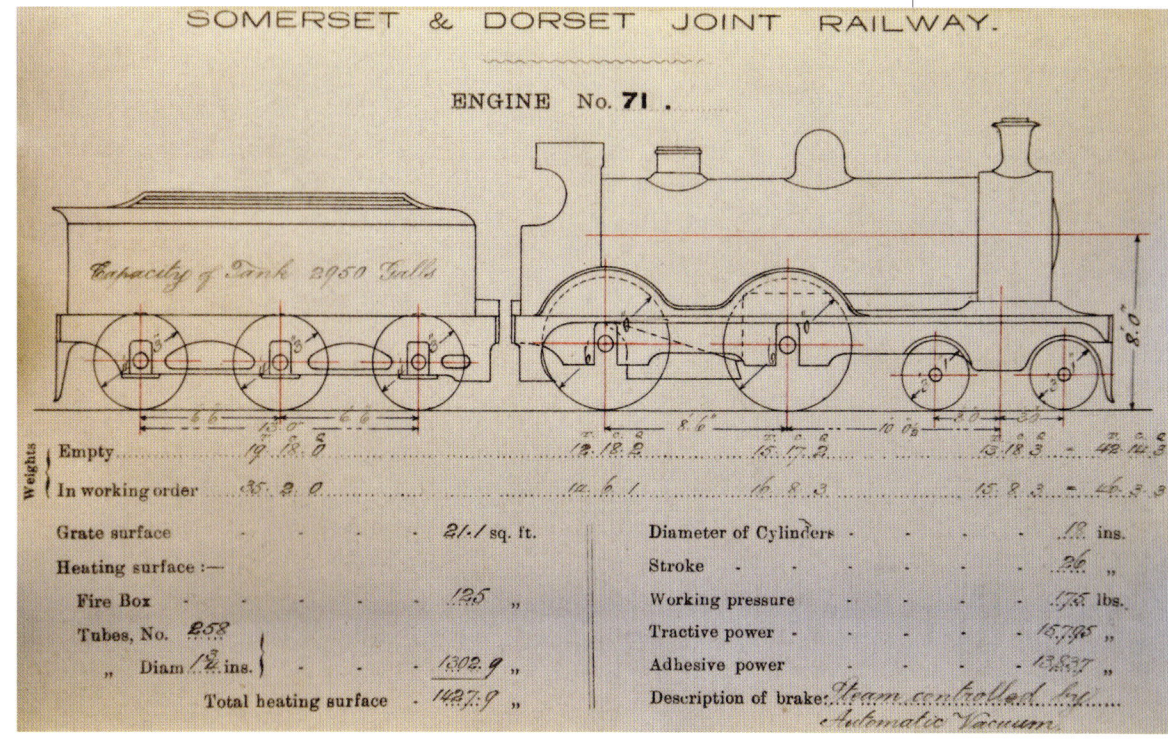

**The three** 4-4-0s that arrived in November 1903 seemed to have sacrificed some of their lean, racy looks for something a little stockier, as these pictures demonstrate. It was an increase in size and power badly needed at that stage by the S&D, as its passenger traffic grew in volume and the weight of new carriages reaching the line from the Midland Railway was increasing. (Author)

**No. 71 at** Bath in about 1910. Appearing as they did at the end of 1903, just before Johnson's retirement on 31 December that year, they were some of the last engines built at Derby to reflect his ideas on locomotive design. They were, as Whitaker requested, larger than the earlier 4-4-0s in a number of crucial areas. Their cylinders were 18in x 26in and the coupled wheels 6ft in diameter. The boiler was six inches longer, its diameter having been increased from 4ft 1in to 4ft 8in., the firebox was 1ft 6in longer at 7ft and contained 258 x 1¾in tubes, which helped produce a total heating surface of 1,420sq ft. and a working pressure of 175lb. In terms of overall loco plus tender weight, they came in at 81 tons and 6 tons, almost 12½ tons heavier than the locos built for the S&D in the 1890s and had a total wheelbase of 21ft 6½in. (D. Neal)

wheel was seized on as a reason for any perceived differences in performance, some found it hard to let it drop. It was an issue that gained additional weight when the smaller engines received 'H' topped round boilers from 1907 onwards and their performance moved closer to that of 69, 70 and 71, but in 1903 this was still far in the future.

While these engines were being built, and perhaps as a swansong to his career, Johnson began developing a three-cylinder compound 4-4-0 version, with a prototype appearing in 1901/02. Four more were soon added and in time the total number would reach forty-five. Some would later describe them as his masterpiece, a creation that evolved into the 1000 Class under Richard Deeley's leadership. Sadly, none of these engines appeared on the S&D even on a temporary basis, but one wonders if these classic engines might have suited the line if the Midland Railway could have been persuaded to part with them?

Johnson's retirement after thirty intriguing years as Locomotive Superintendent could have created a void in the development of the company's motive power fleet. Replacing a man of such skill and standing would be a problem in any walk of life, but particularly in such a creative field. Hamilton-Ellis best summed him up when he wrote:

'He was a personification, a type, of the best locomotive men in the Mid-Victorian and late Victorian years, conservative up to a point but not pig-headed; ready to try innovations, but not make a religion of them; arbitrary when he thought fit, yet genial by nature. He fitted his environment well.'

However, the Midland's managers clearly foresaw his departure and Richard Deeley understudied him during his last year in post, taking over on 1 January 1904 in what appeared to be a seamless transfer of power. It also helped that he had enjoyed a long association with Johnson, beginning in 1875 when he became his pupil. After occupying various posts in the company, he was then appointed Works Manager at Derby in 1902 and Assistant Loco Superintendent the next year. So for many years, he had been able to observe Johnson's work closely and contribute something to his undoubted achievements. In taking over this post, Deeley continued with his predecessor's well established construction programme and gradually bought his not inconsiderable talents to bear on new projects. He was supported in these endeavours by Cecil Paget as Works Manager, who at one stage was rumoured to be his rival for the Superintendent post. With his father, Sir George Paget, as company chairman this was not beyond the bounds of possibility, of course, but Deeley triumphed and went on to grace the post Johnson had done so much to sustain and develop in the post-Kirtley years. However, behind Deeley and Paget was another rising star, Henry Fowler, who had, in 1900, joined the Midland Railway in the dual posts of Manager of the Gas Department and Chief of the Testing Department at Derby. He would, within seven years, rise to the top job and influence the development of the S&D's locomotive fleet for a considerable number of years until retiring in 1933.

In the meantime, Johnson's long career and accomplishments were to be celebrated. He had achieved much in locomotive design and, in the process, had proved to be an effective supporter of the Resident Locomotive Engineers at Highbridge since the Joint Railway came into existence. It had not been a quick process because he couldn't determine the speed of progress; this rested in the hands of the management committee and the company's accountants, with the ever present threat of unmanageable debt foremost in their minds. So he did what he could and supported Fisher, French and Whitaker to the best of his ability, giving them a good, if not perfect fleet of locomotives with which to operate. In retirement, he continued to take an interest in the railways through his membership of the Institution of Mechanical Engineers, where he had been President during 1898. He also remained a strong advocate of electrification, though the scope for such work during his Midland days had been muted. However, it is said that he followed the progress on the Lancaster-Morecambe-Heysham electrified line with interest from his home in Nottingham and witnessed its opening in 1908 first-hand, four years before his death on 15 January 1912. By then he had seen Deeley depart the scene and Fowler take over, witnessing the way they exerted their influence over the Midland Railway and the S&D.

**In many** ways, this photograph sums up Johnson's contribution to the S&D. His 4-4-0 No.15, with its 5ft 9in coupled wheels, and before being reboilered in September 1910, appears to be making light of its load as it passes Branksome shed on the way to Bournemouth in August 1901. It would have been different from Bath to Masbury, where the gradients would see these beautifully proportioned engines struggle with a heavy load if not supported by a second locomotive. (D. Neal)

## Chapter 4
# THE MIDLAND YEARS PART TWO (1904 to 1922)

Richard Mountford Deeley was nearing his 50th birthday when he replaced Johnson as Chief Locomotive Superintendent at Derby. For some people, the weight of succeeding such a man might have proved difficult or impossible to bear. Better, if one must be promoted, to follow someone whose achievements have been modest or insignificant. After Johnson's reign expectations would have been raised much higher and a successor more likely to face a tough, uncompromising scrutiny with little leeway or sympathy if falling short. If he felt such a burden, Deeley doesn't appear to have let it affect his performance and soon began to make his own mark taking Johnson's work to

**A partnership** that was forged in the wake of Johnson's departure and would set the course of locomotive policy on the Midland Railway and the S&D in the years leading up to the Great War. On the left, Richard Deeley as he appeared in the early years of the twentieth century when the phrase 'being in the prime of life' rings true. He was, by any standards, at the peak of his powers and he had the support of the dynamic and ambitious Cecil Paget (right, photographed in 1913) as Works Manager. (Author)

a new level in the process and introducing his own ideas at the same time. However, by this stage he had acquired wide knowledge of his business and extensive experience managing at all levels, often under the most extreme pressure. He was undoubtedly helped in this by a childhood of some affluence and an education well above the norm for the time.

According to his birth certificate, Deeley was born when his family were living in Morleston Street, Derby, on 24 October 1855 within sight and sound of the railway works. His father, also Richard, was at the time an accountant with the Midland Railway, but later moved on to other employment in the farming industry. This included a spell living near Chester which allowed Richard junior to attend Chester Cathedral School from 1868 to 1873, when he left to join the Hydraulic Engineering Company at their works in Charles Street in the city, becoming a pupil of its Managing Director Edward Ellington in the process. From here, he moved to Brotherhood and Hardingham, based in Clerkenwell, London, to continue his training, then returned to Derby to join the Midland Railway in 1875 as Samuel Johnson's pupil. Over the next four years, he worked in the Erecting, Pattern and Fitting Shops before entering the drawing office, where he qualified, then remained as a draughtsman until 1890. During these years, he became an expert in experimental work and testing. This led to his promotion to become Chief of the Testing Department and then Inspector of Boilers from where the path to becoming Works Manager and then Chief in 1904 was set fair. A measure of his growing reputation might be gleaned from his 1906 application to join the Institution of Civil Engineers, which Johnson proposed and was seconded by George Churchward amongst other distinguished engineers of the age.

Meanwhile, Paget had taken a less circuitous route to the top. Born in 1874, he enjoyed a privileged childhood and was then educated at Harrow, before becoming Samuel Johnson's pupil in 1891, a year after his father, Sir George, had become Chairman of the company. His pupillage ran until 1898 and included a three or four-year period at Pembroke College Cambridge where he obtained a science degree. Despite being only 29, he was promoted to become Works Manager in 1903, then General Superintendent in 1907 and so could exert considerable pressure over the way the railway was run and the direction of locomotive policy.

Paget's active and fertile mind constantly sought new, innovative solutions that some thought unrealistic or too risky for any railway company to pursue. His most creative idea centred on the construction of an experimental 2-6-2 tender locomotive which sought to make use of the Willen high speed central valve system which had been designed to generate electric power. It was an idea that had captured his imagination when observing one in operation at Derby and attendance at Pembroke College then gave him the time and opportunity to study it in some depth and gauge its potential as a form of motive power for the MR. The locomotive he created had eight single action inside cylinders with rotary steam distribution. Its boiler was described simply as a 'brick lined furnace with two fire doors at one end and a steel tube at the other, only the steel crown ranked as evaporative surface, for there were no water legs'. In this form, the project developed with some success, but had to be funded from his own resources. When these ran out, the project was handed over to Deeley and was, in due course, quietly dropped, its potential not fully explored or tapped, or so it seems.

However, this was a small but interesting distraction from their primary role, for there were more pressing matters to consider and develop. For example, these focussed heavily on the need to complete Johnson's construction programme and rebuilding other types already in service, more often than not with larger boilers. 4-4-0s continued to be built with another thirty appearing by September 1905, now with water pick-up gear to take advantage of a number of recently installed water troughs. The first ten were fitted with G8 boilers, but with improvements constantly being made, the next twenty engines received the modified G8As, which had a smaller heating surface but could produce a higher psi of 200. At the same time, Deeley continued with Johnson's 4-4-0 compound programme and produced the modified 1000 Class, forty of which rolled out of the works at Derby between 1905 and 1909. While these locomotives were being developed and built, twenty more 0-6-0 goods engines appeared during 1904, followed by another thirty by 1908. They did not ignore the need for

**The size** of the Midland Railway's locomotive fleet during Deeley, Paget and Fowler's time in office can be gauged by this view of the Paint Shop at Derby just before the Great War. It is estimated that as many as 700 engines were painted or re-painted every year, whilst an equal or even greater number were undergoing general repairs lasting many weeks each. To this could be added all the other lesser maintenance tasks undertaken. By comparison, the S&D's needs were small. Nevertheless, through Whitaker's advocacy, the Joint Railway attracted Deeley's attention as it did his successors'. It is rumoured, though not confirmed, that the Chief Locomotive Superintendent was an occasional visitor to Somerset to view operations there. (D. Neal)

more tank engines and constructed forty 2000 Class 0-6-4Ts in 1907, plus five 1528 Class 0-4-0Ts in the same year. Alongside this work was the construction of two steam rail-motors for the Morecombe and Heysham line in 1904 and the development of the electrified line between Lancaster and Heysham that opened in 1908. All in all, it was a busy programme containing an embarrassment of riches that Whitaker probably observed from Highbridge with some envy.

Interestingly, Deeley and Whitaker shared one common idea about the direction in which locomotive development should go. They both saw the need for bigger, more powerful locomotives, though during their time in charge they did not see these ideas come to fruition. In Whitaker's case, his astute analysis of future need had identified a requirement for eight coupled goods engines to meet growing levels of mineral traffic, which he would have observed being developed elsewhere, but particularly by Francis Webb who oversaw construction of 282 between 1892 and 1902 for the LNWR. A little later, John Robinson introduced his version to the Great Central. It appears he may have discussed this with Deeley who, during 1906, produced an outline design, presumably for Whitaker to underline his plans when discussing proposed new engines with the management committee on 10 January 1907.

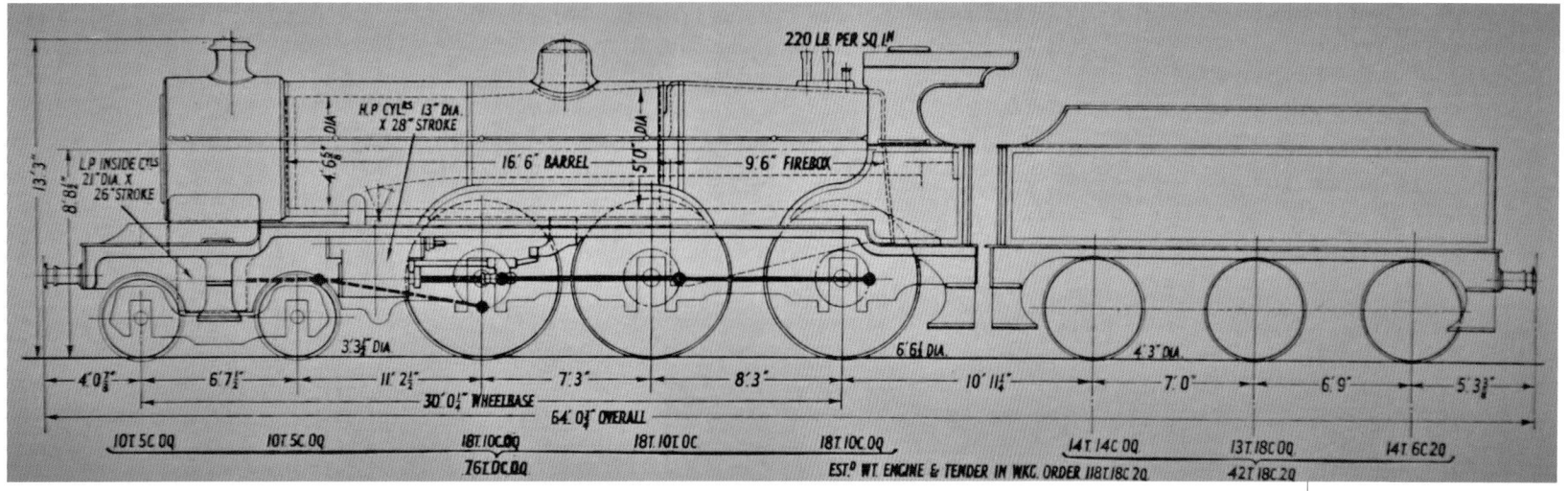

**What might** have been if Deeley's ideas on building bigger engines had come to fruition. Both types would probably have suited the S&D and given them powerful passenger and goods classes that could have coped admirably with the line's idiosyncrasies. (Top) The compound 4-6-0 four-cylinder express engine which would have had 6ft 6½in driving wheels, a total heating surface of 1,970sq ft producing 220 psi and a potential tractive effort of 25,700lb. It has been suggested that this engine wasn't built because Paget's 2-6-2 took precedence, though this is difficult if not impossible to confirm. Equally, it could have been that the Civil Engineer expressed some concern over its weight. Whatever the reason, the MR would not build any engines of this type, though their neighbours, the London and North Western Railway, were happily turning them out from 1905 onwards. These would become a common feature of life when both companies were joined together as part of the 1923 grouping. (Below) The 0-8-0 goods engine, which also got no further than the drawing board, came about in response to a request from Whitaker for six engines of this type. However, in this case the concept evolved into a 2-8-0 design that graced the S&D from 1914 until 1964. (D. Neal)

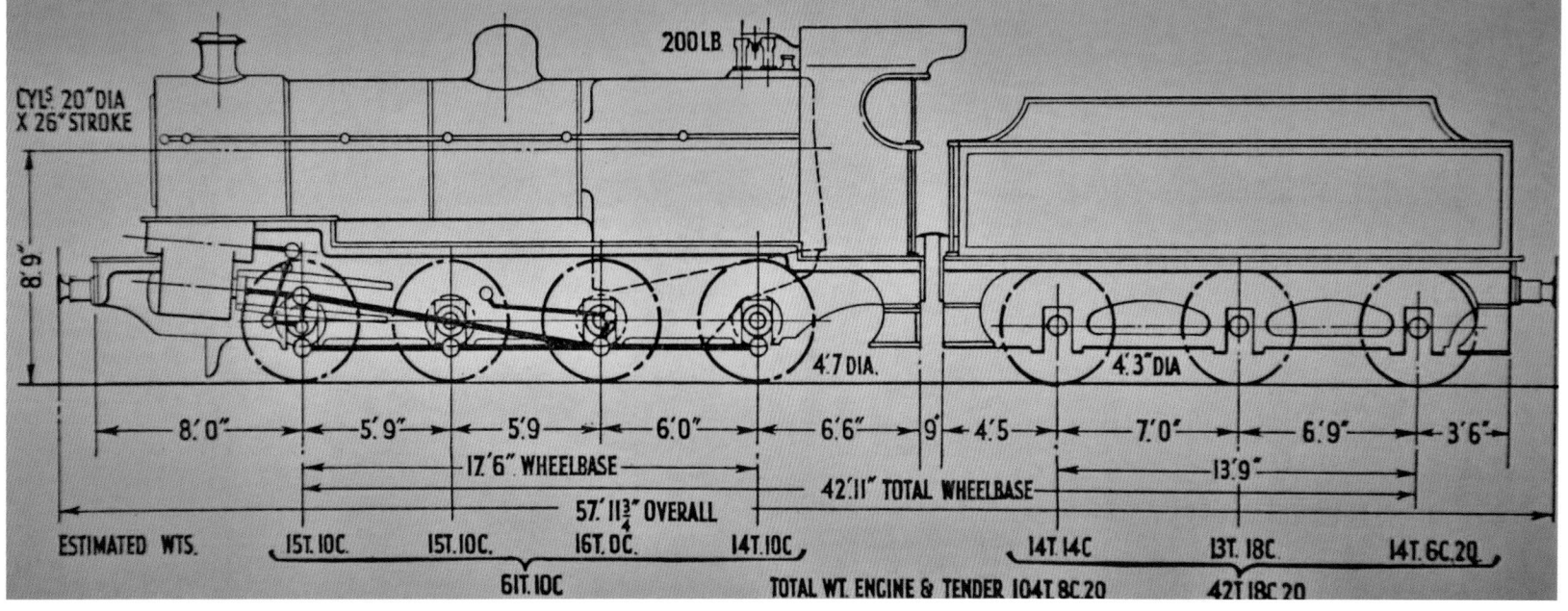

He calculated that six engines of this type would suffice and added three powerful passenger engines, which could possibly have linked to Deeley's plan to build a four-cylinder 4-6-0 compound express locomotive for the MR. The S&D's directors didn't share Whitaker's enthusiasm for this plan, because it meant a commitment to spend £34,700 on strengthening some bridges and sections of track to take these heavier engines. Short-sighted perhaps, but understandable in the circumstances and Whitaker was given an assurance that future infrastructure renewal programmes would seek to address this critical issue. It seems fairly clear, though, that if they had approved his proposal, Deeley's designs would probably have been valuable additions to the S&D where working conditions were hard and favoured bigger engines. It was left to Fowler as Superintendent from 1909 and Chief Draughtsman James Anderson, plus James Clayton and Sandham John Symes, also from the drawing office, to bring one of these ideas to fruition five years after Deeley's departure.

In the meantime, the Midland Railway continued to supply the S&D with another two 4-4-0s and oversaw modifications to other

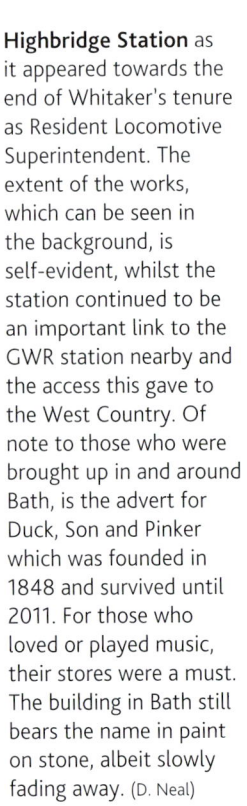

**Highbridge Station** as it appeared towards the end of Whitaker's tenure as Resident Locomotive Superintendent. The extent of the works, which can be seen in the background, is self-evident, whilst the station continued to be an important link to the GWR station nearby and the access this gave to the West Country. Of note to those who were brought up in and around Bath, is the advert for Duck, Son and Pinker which was founded in 1848 and survived until 2011. For those who loved or played music, their stores were a must. The building in Bath still bears the name in paint on stone, albeit slowly fading away. (D. Neal)

**Behind James** Anderson as Chief Draughtsman, then Works Manager and Acting Locomotive Superintendent during the Great War, when Henry Fowler was detached to the Ministry of Munitions, were two designers of great skill. (Left) James Clayton was born in 1872 and became an apprentice with Beyer, Peacock and studied at Manchester Technical School as part of his training. In due course, he became Chief Draughtsman with the Motor Car Company where he met Cecil Paget, who in 1904 employed him as a consultant when designing his 2-6-2 experimental locomotive. Clayton so impressed him that he was recruited to be Assistant Chief Draughtsman on the MR, where he remained until employed by the South Eastern and Chatham Railway in 1914, having failed to succeed Anderson as Chief Draughtsman. (Right) Sandham John Symes was born in County Wicklow in February 1877 and served his apprenticeship with the Great Southern and Western Railway in Dublin where he became a junior draughtsman. After a stint with the North British Railway as a locomotive designer, he was recruited by the MR in 1904 as a draughtsman and was then promoted to become a second Chief Assistant to Anderson, whom he replaced in 1914. Together, they would play a leading role in producing the 7F 2-8-0 class locomotive which, more than any other engine perhaps, came to symbolise the S&D. (Author)

classes of locomotive in the fleet. Engines 77 and 78 were acquired from Derby in 1908, just as Deeley's ideas for a 4-6-0 compound and 0-8-0 heavy goods engine were coming unstuck. The new 4-4-0s followed the pattern set by the three larger engines delivered in 1903 but did contain a few modifications. For example, the H1 boilers contained 242 x 1¾in tubes, sixteen less than their sisters, arranged in vertical rows, which was Deeley's preferred option. However, despite both types of engine producing a tractive power of 15,795lb, Nos. 77 and 78 generated more adhesive power at 14,349lb. In addition, a new style of cab was fitted, the safety valve was modified, as were the handrails, the smokebox doors were secured by clips and bolts, rather than a central wheel, and the chimney was re-profiled. In all other respects, the 1903 and 1908 groups were virtually the same.

Although frustrated in his desire for bigger passenger and goods engines, Whitaker's arguments and Deeley's design work had set a marker for the future, which their successors could develop.

So, by 1908 with a fleet of 13 large and small 4-4-0s, forty-four 0-6-0 tender engines, thirteen 0-4-4s and seventeen other engines of varying types at their disposal, the directors and the management committee could at least argue they now had the numbers to keep the railway running effectively. But times and the world were changing, and this sizeable fleet might not be enough to meet growing demand effectively.

The only other way that Deeley could solve the S&D's perceived motive power problems in the short term was to push through a modification programme, which could hopefully improve the performance of existing engines. This is a natural process in any industry or business and can simply be marked down as good housekeeping. In the case of locomotives, it is a process that fits in with the natural cycle or repairs and periodic maintenance taking place at Highbridge or Derby. For the S&D engines, this rolling programme saw modifications to cabs, new or strengthened frames, new chimneys, the fitting of carriage heating equipment where necessary and tablet exchange equipment to all engines used on the main line. Perhaps most importantly, when it came to performance, there was an active boiler replacement programme. Up to 1904, this tended to be a matter of replacing like with like, because Midland policy still favoured small boilers. But this changed during Deeley's first year in office, when larger boilers began appearing. This change is often credited to him, but in reality, it had its origins in Johnson's last months of service in 1903 and his acceptance of the need for change. However, for the S&D it was a slow-moving programme which only really got underway during 1906/07, there being more

**(Top)** The S&D's 4-4-0 continue to evolve in Deeley's hands with engines Nos. 77 and 78 as portrayed here. However, both he and Whitaker wished for something more than the continuation of this type of engine and hoped that a passenger engine capable of meeting the needs of the company more effectively would soon materialize. Although these engines were an improvement on what had gone before, they were not the solution to Whitaker's problems. With loads increasing, he needed engines that could operate economically and without the added expense of double heading. (Author)

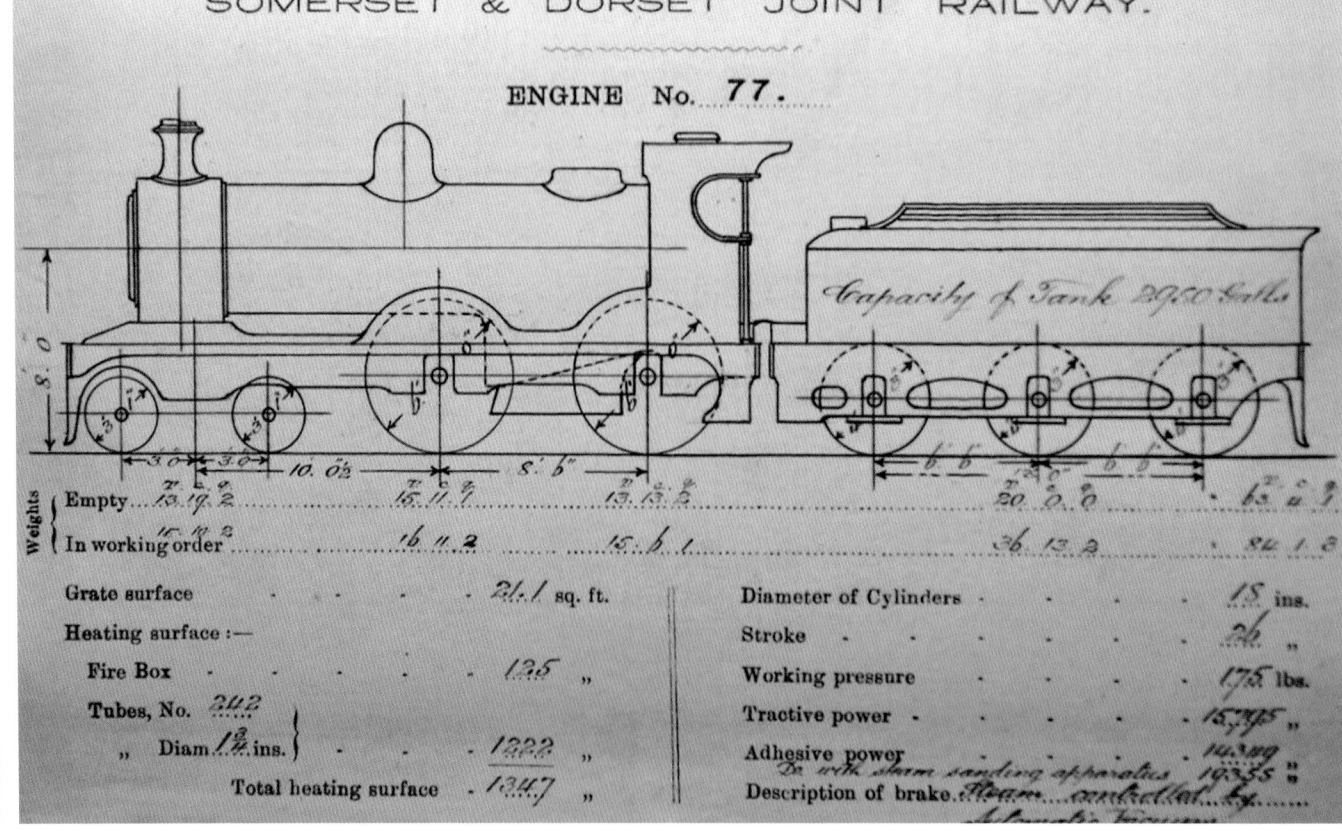

**Under Deeley,** a programme of modifications to existing S&D engines slowly began to change the look and performance of many of its fleet of engines. In many cases, work didn't get underway until 1906/07 then took some years to complete. For example, the first of Johnson's small 4-4-0 tender engines received the bigger 'H' type round topped boiler in 1907 and the last of eight was so fitted in 1911. The engine featured in this photograph, Derby built No. 16, is shown here in modified form at Bath. It was reboilered in April 1906 then received the larger 'H' type in April 1910, with which it remained in service until August 1928. (Author)

pressing demands for new boilers on the Midland Railway.

With so few engines to play with, it was never a massive task for Whitaker and his successors and the workshops. So it was managed in a measured way to fit in with an engine's maintenance cycle. In 1902, this saw the last two 2-4-0s in service, Nos. 15A, ex-19, and 16A, ex-20, built by Vulcan in 1866, being fitted with MR type A boilers, which extended their lives until 1914. This was followed by all nine 1874/76 Fox, Walker built 0-6-0STs, with work beginning in 1906, but not completed until 1911. Then, perhaps most importantly, came eight of Johnson's 'small' 4-4-0s (numbers 14, 15, 16, 17, 18, 45, 67 and 68) which acquired 'H' round topped boilers between 1907 and 1911. This work involved the boilers being shortened by six inches to fit into the frames of these smaller wheel based engines. Three of these boilers (in numbers 17, 45 and 68) were also of the Deeley pattern with vertically positioned tubes, the others being of the horizontal variety. Despite this difference, it is recorded that these modifications enhanced their performance sufficiently to compare favourably, in power and performance, to the bigger 4-4-0s. True or not, it seems that these modified engines were favoured by footplate crew over the newer, bigger engines for their strength and fuel economy. Whether large or small, four of these engines would be involved in one last one round of boiler upgrades. Between 1921 and 1927, Nos. 17, 45, 77 and 78 would receive G7 Belpaire boilers which would see them into the 1930s, when they were withdrawn from service.

Other than modifying the S&D's existing fleet where possible and seeking to build 4-6-0 and 0-8-0 types of locomotive in the face of strong opposition, Deeley was limited in what he could do for Whitaker. However, for most of his time in charge at Derby, limitations were placed on what he could achieve by the S&D's own management committee, ever aware of the company's lack of financial clout. And there was also the question of resistance to his ideas emanating from the Midland Railway's own senior managers, led by Sir George Paget, the chairman, and Guy Granet, the General Manager. His position was probably not helped by Cecil Paget's promotion from Works Manager, and Deeley's deputy, to become General Superintendent, and his superior, or by Granet being succeeded by Henry Fowler in 1907. George Paget, who remained Chairman until 1911, may have been instrumental in this change and the charge of nepotism did hover in the air for a time. However, Paget junior was a man of intelligence and drive who had a very clear view of the direction in which the company should move.

**Two more** modified Johnson 4-4-0s busying themselves around Bath. Engine No. 15 was built in 1891 at Derby, but there is some confusion over reboilering dates as there is with its withdrawal from service. One source lists these as – reboilered for the first time in December 1904, received 'H' type in September 1910 and was withdrawn in August 1928. However, the National Railway Museum records in its collection that the dates were April 1905, June 1911 and September 1931 respectively. A small but intriguing difference that shows how difficult it can be to set out a completely accurate picture at times. For engine No. 18, formerly No. 45, only one set of dates survives. An 'H' type boiler was fitted in August 1909 and a G7 Belpaire in September 1926, with which it is seen in this photograph taken in about 1928, four or so years before withdrawal from service. (Author)

So, this promotion, whilst coloured by a suspicion of favouritism, was not undeserved or as risky as it might seem at first sight. But seeing one's junior promoted over one's head can be a difficult pill to swallow and in 1909, differences of opinion, and the rejection of Deeley's ideas on locomotive policy, brought matters to a head. The spark for this seems to have been a re-organisation that Granet approved that would see the Chief Locomotive Superintendent post divided into two, with the creation of a Chief Mechanical Engineer and a Chief Motive Power Superintendent. Unable to accept this change, and with the rejection of his development work on larger engines still fresh in the mind, Deeley chose to resign; a decision sweetened by the award of a substantial pension. And with this he departed the scene on 13 August, to spend the rest of his life researching areas of science that captured his imagination, including meteorology and the development of lubricants. He died on 19 June 1944 at the age of 90 at home in Isleworth, West London.

Deeley's successor was his deputy and Works Manager, Henry Fowler, who seems to have been the only candidate considered for the post. Nevertheless, he was well qualified and had by 1909 acquired considerable experience of locomotive design and construction. Born in Evesham on 29 July 1870, he attended Mason Science College (now part of the University of Birmingham) on leaving school, where he studied for an Engineering Diploma. His work became focussed on the physical and chemical behaviour of metals

**The S&DJR's** reboilering programme was a continuous process during Deeley and Henry Fowler's time in charge as Chief Locomotive Superintendent and Chief Mechanical Engineer respectively and included most classes operated by Highbridge. However, in some cases this wasn't a major modification affecting all members of a class but simply a matter of routine maintenance – an old boiler being replaced by one of the same type that was new or refurbished. There were exceptions, though. (Top) Only three 0-4-4Ts received G6/G5 Belpaire boilers (Nos. 32, 53 and 55) as witnessed here by engine No. 55 which was so fitted in 1925. In 1905/06, it had simply been reboilered on a like for like basis. The fact that only three of the class were modified in this way suggests the expected benefits weren't substantial enough to justify the additional expenditure. However, No.32 did last longer in service than any member of the class (until 1946, fourteen years longer than any other). (Below) The 0-6-0 tender engines didn't begin receiving Belpaires until the early 1920s and then over a period of ten years fifteen received either G5 or G7 types, whilst during 1910, Derby built engines Nos. 63 and 66 were fitted with 'H' round topped boilers. Both of which were replaced by G7 Belpaires in 1920. In this picture, engine No. 64 can be seen after it received its G7 in March 1921. (D. Neal)

**In the** pre-war years, before a new reality descended and Britain changed beyond all recognition in a social and economic way, the sight of a Vulcan built 2-4-0, in this case No. 15A, passing through countryside that surrounded the S&D, conjured up an evocative picture of a world soon to disappear. (D. Neal)

and the application of metallurgy to engineering. In this he was helped considerably by studying under Professor Thomas Turner, a pioneer in this area of research. Turner became so famous in this field that during the 1930s, he would be employed as consultant by Nigel Gresley in the design of the LNER's locomotives and his son, Thomas Henry Turner, would become Gresley's Chief Chemist and Metallurgist at Doncaster. Fowler maintained a close interest in this subject throughout his life and this clearly influenced his work with the MR and then the London, Midland and Scottish Railway.

Although educated to an advanced level in science, there will always be a need to find a practical outlet for this knowledge. This he found by becoming John Aspinall's apprentice pupil with the Lancashire and Yorkshire Railway at its works in Horwich, to the West of Bolton. Aspinall was a key figure in the railway industry at this time and an inventor of some note who would in time have fourteen patents to his name. The list of locomotives to his credit was also impressive – 484 Class 27 0-6-0s, 270 Class 5 2-4-2Ts, 70 Class 2 and 3 4-4-0s, 60 Class 30 0-8-0s and more. With such a background he was the ideal man to take Fowler in hand at this most crucial stage of his career. And so, between 1887 and 1891, the young man broadened his range of skills before transferring to the Testing Department of which he later became head. Here his knowledge of metallurgy was put to good use and this helped enhance his growing reputation as an engineer and manager. Such was the impact he made, that promotion to Gas Engineer followed and during 1900, in an attempt to further his career, he joined the Midland Railway, rising to Assistant Works Manager before succeeding Paget in 1907 and Deeley two years later.

When becoming CME, he inherited a large and varied fleet of locomotives, but one based upon a small engine policy to which the company had become wedded. As we have seen, Deeley sought to widen this brief in the certain knowledge that larger, more powerful locomotives were the

**Despite the** progress made during Johnson and Deeley's years in charge, some of the original 1860s locomotives still struggled on finding ready use into the 1920s. This was particularly so with engine No. 7, which was converted from a 2-4-0 tender engine to a 2-4-0T in 1888 having been renumbered 27 then 27A in 1876 and 1881 respectively. In this form, it is seen here post-Great War, in which condition it would survive until July 1925. Some of the men who would be working locomotives until the line closed cut their teeth on this engine and its other surviving sister No. 28A which lasted until 1928. (Author)

way ahead. He saw that loads were increasing, and the company were relying too heavily on the expensive business of double-heading when a single bigger locomotive might suffice. The GWR under George Churchward, for example, had embraced this change early in the century. As a result, they began producing their impressive Saint Class 4-6-0s in 1905, building 77 by 1913. To this they added eighty-four 2-8-0 Class 2800 heavy goods engines between 1903 and 1919 and a single experimental Pacific in 1908. And from this promising start, other types would soon evolve in the works at Swindon. Other companies were slower to respond to this changed perception, though embraced it more fully in the post-war years as grouping became a reality. For Fowler, this was still far in the future when becoming CME and most of the projects he was allowed to develop before the war could, for the most part, be seen to follow the MR's dictates on size. However, there would be one exception to this – a 2-8-0 class engine for the S&D.

With Fowler in charge, Anderson became Works Manager and Symes was promoted to Chief Draughtsman. He and Clayton were both candidates for this post and there was a natural expectation, due to his greater experience and seniority, that Clayton would be appointed and not Symes. Finding his promotion path blocked and, perhaps, still unable to accept this apparent snub, he departed the scene in 1914, finding employment with the South Eastern and Chatham Railway as Leading

**As the** Great War approached, the works at Highbridge had become a much busier place as witnessed by this photograph. This isn't surprising, though, because during Whitaker's time not only had the number of locomotives increased appreciably, but also the works had been enlarged and improved to absorb the extra work. (D. Neal)

Locomotive Draughtsman. He was soon promoted to Chief Draughtsman and then became the CME's Personal Assistant, transferring with Richard Maunsell to the Southern Railway when it was formed in 1923. In this role, he led the way in the construction of many Maunsell designs, including the King Arthur Class and the Lord Nelsons. In so doing, he demonstrated skills that the Midland Railway, and then the LMS, might have found extremely useful. However, before leaving Derby he was deeply involved in Fowler's pre-war locomotive projects, three of which affected the S&D.

However, before this work could reach fruition, Fowler authorised a trial on the S&D using a Deeley designed 0-6-4T. This exercise was arranged to see if this class might meet Whitaker's often repeated request for a more powerful locomotive to service the increasingly heavy loads during the summer months. Why this class of engine was chosen is unclear. The forty built in 1907 proved to have poor acceleration and were found unstable by their crew; two issues that would eventually see them relegated to freight duties following two serious derailments in 1928. One can only assume that,

as they were related to the 0-4-4Ts, which had found ready use on the S&D, someone at Derby thought these powerful engines, with their 19,756lbs of tractive effort, might find similar use on their metals. Alternatively, they had proved to be of dubious value to the Midland Railway and another use was being sought in an effort to offload them. Either way, the problems uncovered at Derby were quickly confirmed by Whitaker's staff and the sole engine, No. 2023, was returned after a number of trial runs had soon exposed its weakness. In fact, this proved to be one of Whitaker's last acts as Resident Locomotive

**No. 2000,** the first of Deeley's 1907 built 0-6-4Ts. As a class, they acquired the nickname 'Flat Iron' in reference to their side tanks which extended to the front of the firebox. They were developed from the Johnson 0-4-4Ts that had become so familiar to the S&D, but the extra set of driving wheels and their greater strength did not make them better engines. When No. 2023 was sent to the S&D for trials in April 1911, its lack of power, poor stability and restricted tank capacity were quickly exposed and it was soon sent home. The fitting of a Belpaire firebox and superheating between 1920 and 1926 seems to have done little to improve them and they saw out their days on freight duties with the LMS, the last being scrapped in 1938. (Author)

Superintendent. Three months later, on 24 July, he retired, making way for Mervyn Ryan, who was promoted into the post from Derby.

Ryan was a highly skilled engineer by the time he reached Highbridge and was probably over-qualified for the post, which by any standards was something of a backwater when compared to the cut and thrust of Derby. Born in Malta in 1883, he attended the prestigious Stonyhurst College in Lancashire. When choosing a career, he followed a well-worn path into the railway industry and began a five-year pupil apprenticeship under Deeley and Fowler at Derby in 1902. At the same time, he enrolled at University College Nottingham where he specialised in Civil Engineering. After spending time in a number of workshops and occupying the position of Pupil Inspector for a while, he was seconded to the Schenectady Railroad in New York in 1906 and the Pennsylvania Railroad at Altoona the following year. In both places, he was attached to the Running Shed and also undertook firing duties, but in New York he also studied electrical engineering for some months. Railway electrification projects in the city were by then being considered and by the end of the decade would become reality.

On completing his apprenticeship, in 1907 he was appointed Inspector of Stores at Derby, then Assistant Works Manager two years later, from where he was appointed to Highbridge. It isn't known if he sought this posting, seeing it as a step upwards, or took it as a favour to Fowler, who may have wanted someone with his skill and reputation to follow the well-established Whitaker. Either way, he didn't remain long with the S&D. In 1913, he resigned and took the post of Assistant Works Manager with the London and South Western Railway at Eastleigh working for the CME, Robert Urie. During the

**A sight** that became familiar to Mervyn Ryan during 1907 when he was attached to the Pennsylvania Railroad at Altoona for 'running shed and firing duties'. He remained with the company for many months during which time he would have become aware of their huge number of 2-8-0 locomotives including this H Class engine, photographed with its crew and other workers. Of equal interest to Ryan and other locomotive engineers world-wide would have been Altoona's locomotive testing centre that opened in 1875 and was fitted with a static testing facility in 1905, in the same year that Churchward opened his own rolling road at Swindon. (Author)

Great War, he was re-united with Fowler at the Ministry of Munitions and was promoted to become Director of Munition Gauges. He returned to the LSWR, now with a CBE for his war service, but for some reason decided to move on and was appointed CME of the Argentine Railway in 1919, dying there in the British Hospital in Buenos Aires on 22 April 1952 from cancer. A measure of the respect in which he was held may be gained from his application to join the Institution of Civil Engineers in 1910. Henry Fowler was proposer with both Johnson and Deeley acting as supporters and referees, a great accolade indeed from three such talented engineers. Then, in 1919, he was elected President of the Institution of Locomotive Engineers, another sign that his star was in the ascendency. With these achievements behind him, and more to come, it is likely that he would have risen to become the CME of one of the Big Four companies created in 1923. But this wasn't to be, and Britain's loss became Argentina's gain. However, for the purpose of this story it is his two years with the S&D that are most important. It was during this very short period that his skills and actions helped realise Whitaker's long held ambition to acquire a class of heavy goods locomotives for the company.

Before taking over this post, Ryan probably observed the debate initiated by Deeley and Whitaker over the need for bigger engines, in particular an 0-8-0 heavy goods locomotive. And one of his first acts on arriving at Highbridge, possibly encouraged by Fowler, was to look at this failed proposal with a fresh pair of eyes. He soon discovered that the case made three or so years earlier, and rejected on cost grounds, could be re-balanced and made more attractive to the management committee as well as the civil engineering department. Faced with a substantial bill to upgrade bridges and track to take eight-coupled locomotives weighing up to 60½ tons, to which might be added the cost of a new turntable at Evercreech Junction, they had baulked. But Ryan demonstrated that by going for an engine with a pony truck would mean 6 tons less weight carried by the eight coupled wheels. This simple measure would, he surmised, reduce the civil engineering bill by £32,650 to £2,050. The turntable issue could, he believed, be resolved by having a locomotive fitted with a tender cab allowing it to

**Mervyn Frederick** Ryan, when elected President of the Institution of Locomotive Engineers in 1919, a position held by Fowler, Stanier, Gresley, Bulleid and other famous engineers. After a rapid rise through the ranks, and with the potential to become a very senior figure in Britain's railway industry, he was attracted by a new challenge in Argentina. Although Locomotive Superintendent at Highbridge for a comparatively short period, he astutely revived Whitaker's plan for larger freight locomotives to cope with growing loads and the S&D's challenging gradients. His analysis and reasoned arguments ensured six 2-8-0s were built and the railway received rebuilt 483 Class superheated 4-4-0s as well. It is interesting to speculate how far this talented engineer could have changed the locomotive fleet and the workshops if he had been persuaded to stay at Highbridge for longer. (Author)

**One of** more than 2,000 Class H6 2-8-0s built for the Pennsylvania Railroad between 1899 and 1913. During his secondment to this company in 1907, Ryan would undoubtedly have observed these engines in action and may even have worked on the footplate, this being the primary reason for his visit to the USA. More than 1,500 H8/9/10s would be added to this substantial fleet by 1919. Did these PRR engines capture Ryan's imagination and encourage him to explore their potential when promoted to Locomotive Superintendent of the S&D? However, there were a number of 2-8-0s making their appearance in Britain that could have been of equal importance to his work. (Author)

work backwards when necessary. Armed with this new evidence, he re-assessed the likely day to day savings achievable by operating six 2-8-0 engines and demonstrated a potential benefit of £8,650 per year if they replaced eight older locomotives; a calculation that assumed that heavier loads could be carried by the bigger engines. Ryan's arguments carried the day and Fowler was authorised to build these six engines at a cost of £3,500 each.

By this stage, the concept of engines with this wheel configuration was still fairly new in Great Britain, but in the USA, it had been in various states of development for more than forty years. The first 2-8-0 tender engine made its appearance on the Lehigh and Mahanoy Railroad in 1866, having been built for them at the Baldwin Works in Philadelphia. Given the generic name 'Consolidation', they gradually grew in type and number. In fact, Ryan would probably have observed them operating on the Pennsylvania Railroad when attached to this company in 1907. They had acquired their first example of this type in 1875 and would in time make it their standard freight engine until 2-10-0s were introduced in 1916. By the early twentieth century, the numbers were still multiplying and had passed through various marks from H1 to H6, with more to follow

**Ryan didn't** have to look far for other examples of 2-8-0s being constructed in Britain during the early years of the twentieth century, as witnessed by this Churchward Class 2800 locomotive, No. 2864, pictured at Swindon, and in diagram form. Eighty-four of these engines were built between 1903 and 1919 and led to the Class 2884 which first appeared in 1938. The prototype, No. 97, (later renumbered 2800) underwent two years of trials before the class went into full production. From 1909, superheating was fitted to these engines and in this state they continued to serve the railway until the 1960s, so successful was the design. (D. Neal)

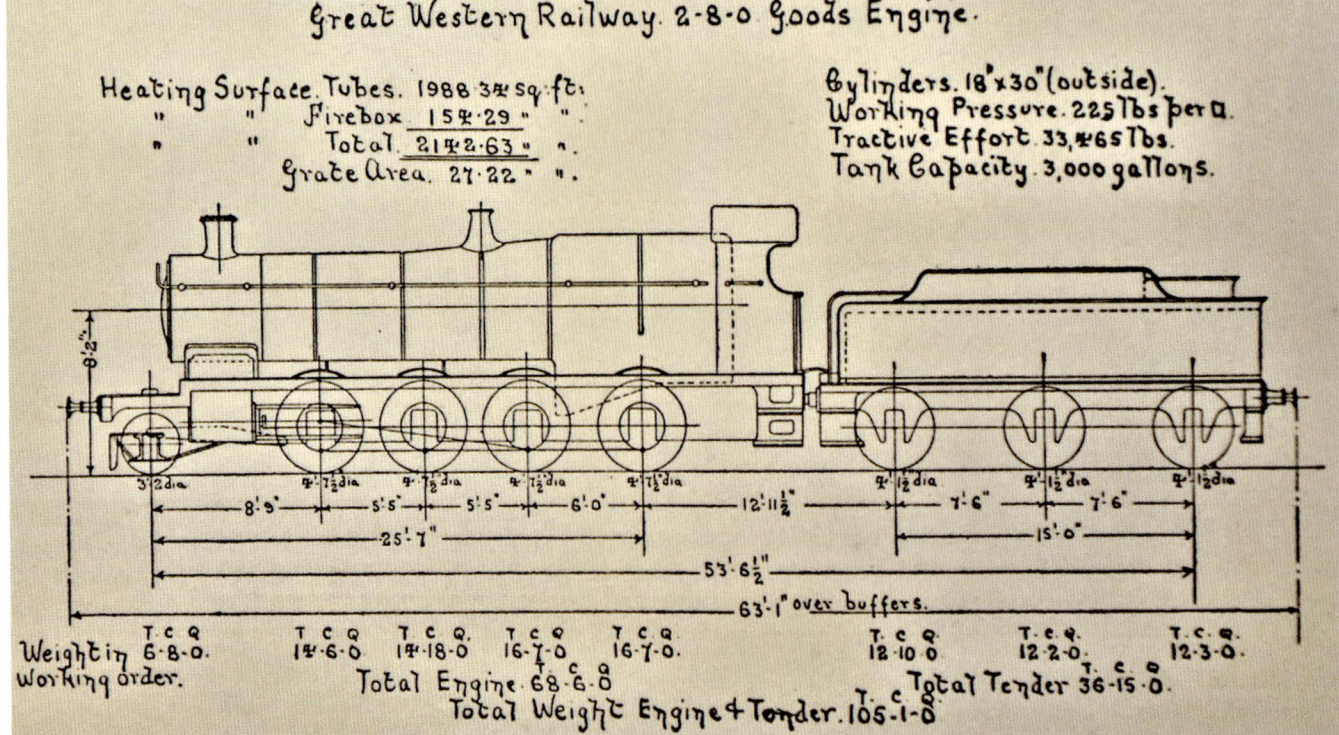

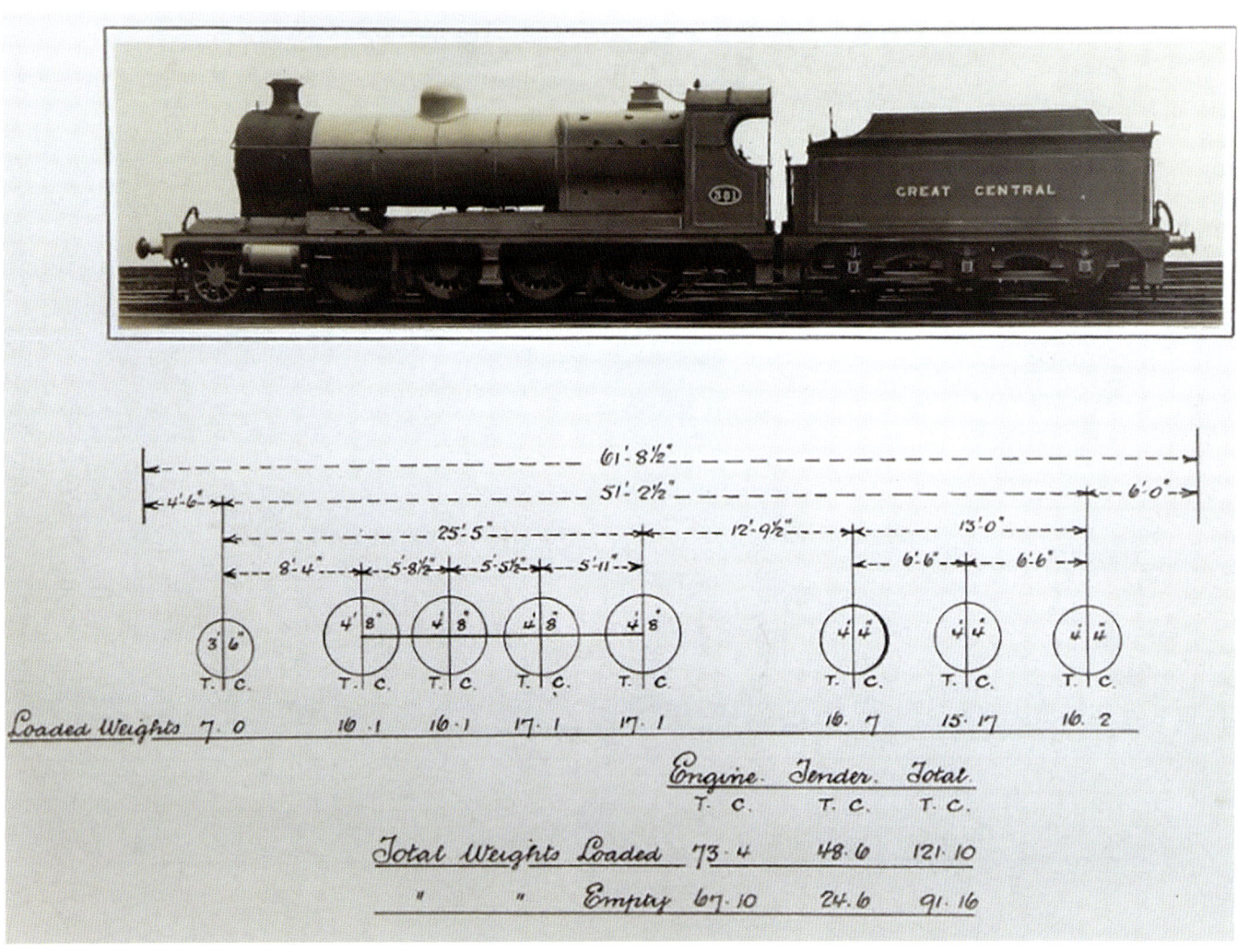

**Next came** John Robinson's Great Central Railway Class 8K which began appearing in 1911, with 126 being built by 1914 when the first S&D 2-8-0 entered service. This class was selected for development by the Railway Operating Division that controlled the supply of locomotives to the Army overseas during the Great War; 521 were built for this purpose. In service, they all proved very effective, being noted for their strength, reliability and good steaming qualities. (Author)

until the H10 appeared during 1916. By the time Ryan visited the States, the most numerous of these classes were the H6, of which 2,034 were built.

When considering the S&D's requirement for these 2-8-0s, the draughtsman at Derby, with Ryan hovering in the background until his departure in September 1913, had three British examples to observe, each being described in great detail in the railway press at the time. Churchward's Class 2800 were by then well-established and Ryan no doubt observed them at close quarters from his office at Highbridge as they passed down the GWR's main line close by. Then there was the Great Central's more numerous 8Ks, all 126 of which were in service by 1914. Meanwhile, the first five of Gresley's O1s were being built at Doncaster in 1913/14 for the Great Northern. An embarrassment of riches for Anderson, Symes and Clayton to study as their own plans for six 2-8-0s came to fruition. And it was to Clayton that the work seems to have devolved as his time to leave Derby for pastures new approached, if the few sources of information still available are to be believed.

While this was going on, Fowler, probably with Ryan's agreement or perhaps even at his instigation, authorised the transfer to the S&D of a 483 Class 4-4-0 superheated locomotive in the spring of 1913 for trials purposes. The aim was a simple one – to compare its performance to the larger 4-4-0s Derby supplied in 1903 and

**The third** type of 2-8-0 that may have influenced Ryan and the designers at Derby was the Gresley O1 class that entered service with the Great Northern Railway in 1913. These engines were built with two cylinders, not unusual for Gresley at this time, but he would soon begin developing his three cylinder locomotives with 2 to 1 motion. To this end, the next 2-8-0 to appear, No.461 during 1918, was a prototype to test his theories and from this his most famous classes of locomotives began to evolve. The engine photographed above, No. 456, is the first member of the O1 class. (Author)

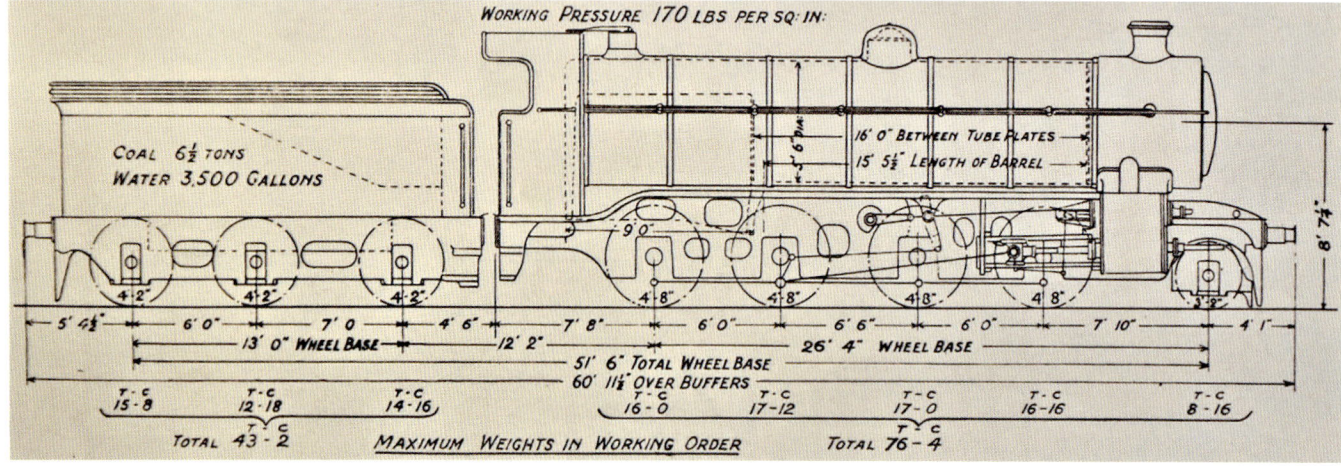

1908. To allow these trials to be developed, No. 499 was soon joined by sister engine No. 519 and through the summer they both worked many of the heaviest express services. These two engines, with 158 others, were built between 1882 and 1901 to a Johnson design, but under Fowler were rebuilt. They retained their Stephenson valve gear but, amongst other things, they acquired superheated boilers, 6ft 5½in driving wheels and 18in by 26in cylinders with slide valves. Their conversion began in 1910 and by 1913 sufficient numbers were available for these trials on the S&D to begin.

Superheating began to come into its own in the late nineteenth century, building on development work undertaken in Germany by Doctor Wilhelm Schmidt, much later with the assistance of Doctor Robert Garbe, CME of the Berlin Division of the Prussian State Railway, and his opposite number at the Belgian Railway, Jean Baptist Flamme. The aim of this research was to produce a system that converted saturated steam into superheated steam, so increasing a locomotive's thermal efficiency. In so doing, the consumption of water and coal were reduced, though there was an additional cost in construction and maintenance to be considered. However, in practice this did not appear to outweigh the benefits accrued from improved performance or any savings achieved in lowering rates of consumption.

The first engine so equipped is believed to be the Prussian Railway's 4-4-0 Class S4, which first appeared in 1902 and by 1909, 104 had been built. Initially,

**The first** 483 Class 4-4-0 locomotive as it would have appeared in the first decade or so of the twentieth century. It was in this state that two others of the class, Nos. 499 and 519, operated on S&D metals in 1913. (Author)

**Three 4-4-0 class** locomotives on duty at New Street in Birmingham, two of them 483s. Engine No. 527, which is on the turntable and equipped for oil firing, is of interest because this engine is recorded as being part of the S&D's fleet post-Second World War and was withdrawn in February 1956 from Bath Shed and scrapped at Derby a month later. The other 483 Class engine is apparently No. 539, whilst the engine nearest is unidentified but still has a round-topped boiler. (D. Neal)

**In 1913** two superheated 483 Class locomotives were loaned to the S&D during the summer months for evaluation. As a result, five of their 4-4-0s were withdrawn for conversion between 1914 and 1921 – three large (Nos. 69, 70 and 71) and two small (Nos. 67 and 68) – and replaced by modified engines. The picture above captures the 'new' No. 67 as it appeared during the 1920s. At the time, they were reported as having high pitched G7S boilers, Schmidt type superheaters, a Fowler style chimney, extended smokebox, 7ft coupled wheels, 20½in x 26in cylinders, Stephenson valve gear, bogie brakes and more. The report then confirmed them as being Class 4P/2G engines and added the title 'Standard MR Class 2 4-4-0s'. In this condition, these five engines saw service into the 1950s. (D. Neal)

they were fitted with smokebox superheaters, but these were only found to be passably effective but no more. So a smoke tube version was developed and from 1906, these were fitted to the S4s. Ever aware of scientific developments around the world that might have benefits in his own area of expertise, George Churchward observed Schmidt's work and the progress being made in Germany. In response, he set about producing a GWR version and in 1909 came up with a workable solution. These featured, most notably, in his Saint Class 4-6-0, which, between 1909 and 1912, were fitted with the Swindon Superheater No. 3. Meanwhile, Robinson on the Great Central was also developing his own system which was attached with some success to the GCR 8K 2-8-0s from 1911. In due course, Robert Urie, noting these advances, fitted a mixture of Schmidt and Robinson superheaters to a number of his H15 4-6-0s in 1914 as an experiment to see which was more effective. Finding both falling short of his expectations, he and his team set about designing their own version which then found wider use on LSWR engines.

So the spread of superheating continued and Johnson's 483 Class 4-4-0s were duly modified with Schmidt superheated G7 boilers and Belpaire fireboxes, with engines Nos. 499 and 519 proving their worth to the S&D during 1913. In October, after a successful summer season, they were returned to Derby. Soon instructions were given to withdraw engines Nos. 69 and 70 from Highbridge's fleet for rebuilding. In fact, No. 71 replaced 69 in this schedule, its frames

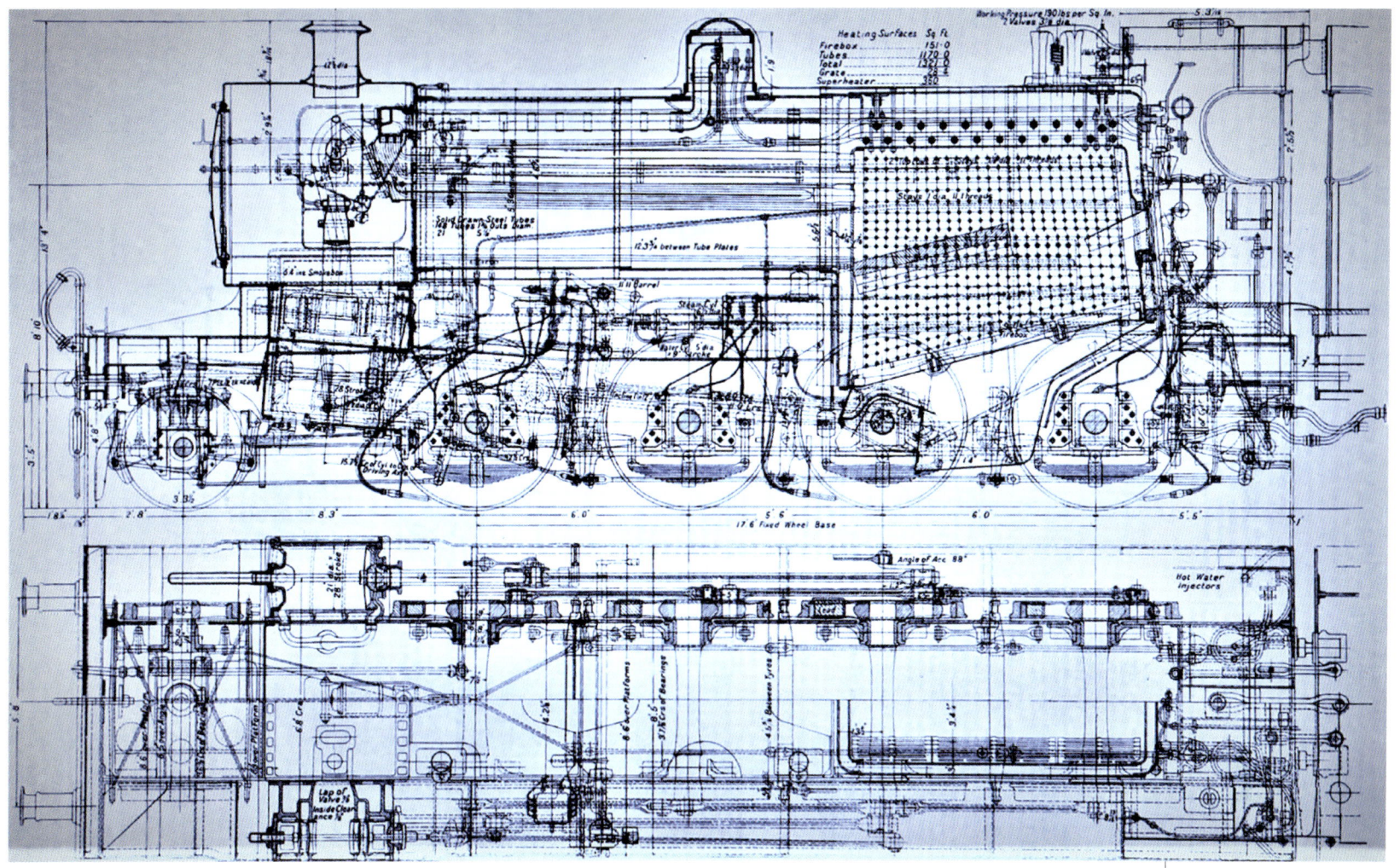

being discovered to be in very poor condition. In due course, the old boilers from these two engines were attached to new fireboxes and then fitted to S&D 0-6-0s Nos. 63 and 66 during 1914, receiving new frames at the same time. This, by all accounts, boosted their performance considerably, making them ideal substitutes for express trains during the summer months. In the meantime, two rebuilt 483 Class locomotives were allocated to Highbridge in the spring of 1914 and given the numbers 70 and 71. Engine 69 was due to be replaced in 1916, but the war delayed this programme and a new 69 didn't arrive until April 1921. At the same time, two of the smaller 1896 Johnson engines, Nos. 67 and 68, were taken out of service for conversion to 483s and replaced, upping the S&D's allocation to five superheated engines. As this programme slowly evolved, Fowler's 2-8-0s became recipients of this developing technology.

It seems that surviving drawing office and workshop records describing the creation of these engines are incomplete, so making it difficult to establish and describe the complex route that all engineering projects of this size must take. Who did what and why remain hazy issues, as does the discussions that took place as proposals evolved and were approved or rejected. Although there would have been a specification for the project,

**The design** for the S&D's new 2-8-0 is captured in this early diagram prepared for publication in the railway press. In designing this locomotive, Fowler and his drawing office team at Derby produced a compact and powerful engine, more than capable of meeting the company's needs. (Author)

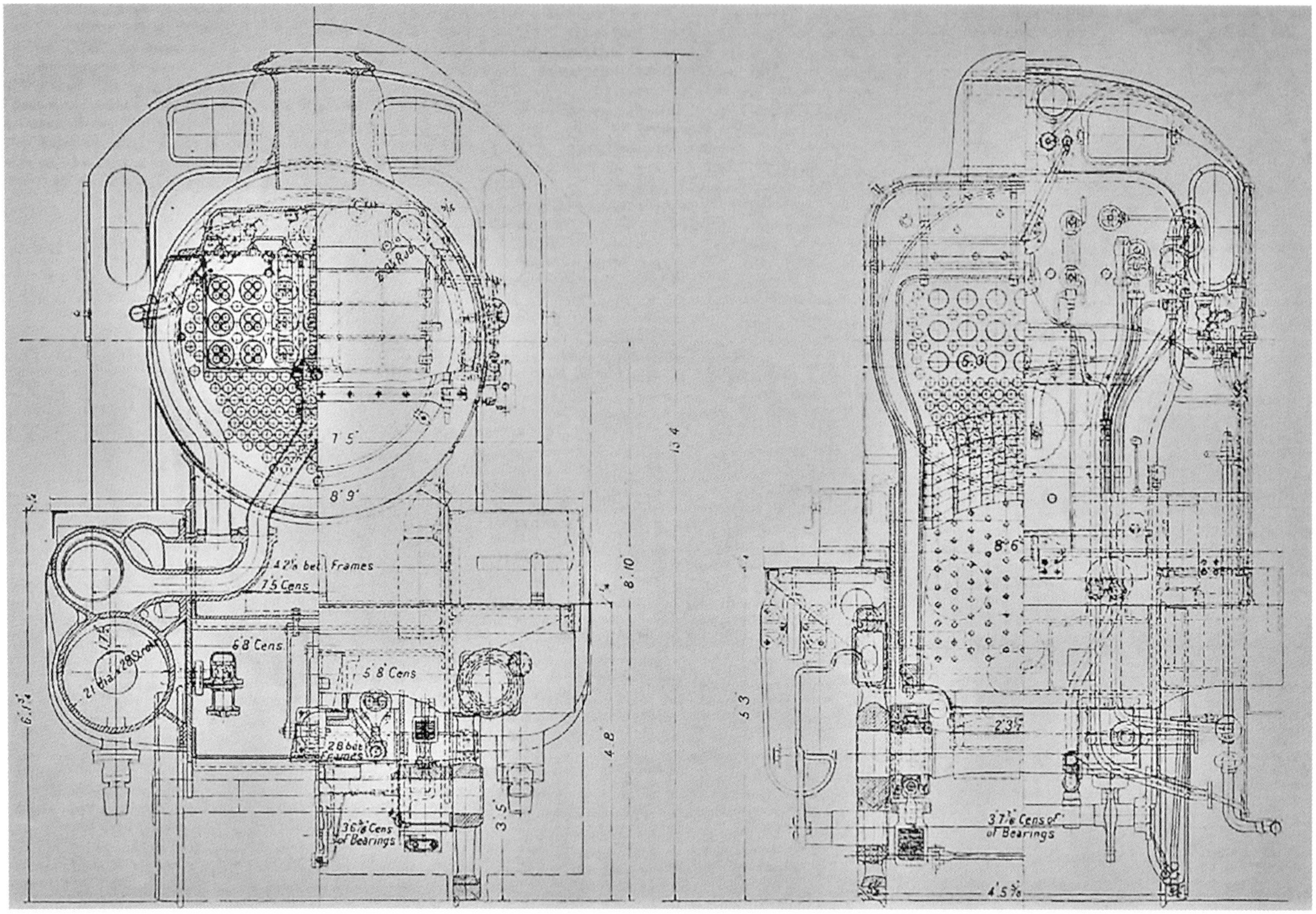

**Although faded** with age, these drawings still reveal some aspects of the 2-8-0s' design, in particular the cylinder and boiler arrangement. Forward visibility from the cab, never a strong point with many steam locomotives, seems in this case to be quite good. Without a turntable capable of taking these engines at Bath, until the mid-1930s, tender first working became an essential part of these engines' lives. The tender cab, although designed to provide protection from the weather, did also hinder the rearwards view. Deemed by footplate crew to be unnecessary, they were removed between 1918 and 1920. (Author)

**The first** S&D 7F 2-8-0 (as they became later under LMS ownership) makes its appearance having passed through the paint shop at Derby early in 1914. The locomotive has been grey 'primered', and awaits its number, in this case 80, and the company insignia to be painted on its tender. (Author)

there would still have been much for the draughtsmen to mull over and resolve. Being the first engine of this type built at Derby, the designers would have looked at different options and sought to balance these in an effort to meet the specific requirements of the S&D. And there were many variables to consider – wheel size, boiler dimensions and capacity, grate area, type of superheater to be installed, cylinder size and valve gear and much more. By this means, and through continuous negotiation with Ryan and senior managers at Derby, plans were gradually refined and given final approval. When complete, the first steps in production could begin with the ordering of parts from suppliers and raw materials which would then be forged and machined as necessary. Even when assembly was underway, refinements would continue to be made to the design, so enabling it to evolve in the light of slowly acquired experience.

The locomotive that was gradually put together did draw some elements from the other 2-8-0s produced in Britain since Churchward's Class 2800 first appeared in 1903 but was certainly no copy as its looks will quickly bear witness. It shared cylinder size and Walschaerts valve gear with Gresley's O1s, while other components, such as wheels, were similar in dimension and all had six wheeled tenders, though water and coal capacity varied slightly.

A press report written in 1914 coined the words 'simple designs in many ways making them easier to service and maintain, but sturdy reliable engines nonetheless' when describing all four classes. In terms of comparable performance, it would take a while to establish a pecking order, but with a tractive effort of 35,295lb (though sometimes quoted as 35,392lb at 85 per cent boiler pressure) it was better than either Robinson and Gresley's engines and only slightly below the 2800s at 35,800lb. However, the S&D's engines were substantially lighter than the 2800s, O1s and 8Ks by nearly 4 tons, 12 tons and 9 tons respectively. This produced an axle loading of 16 tons for the 7Fs which may

**The first** engine is again on display but now numbered and company identified, but still in grey primer apparently with some lining. As was the custom, it will be photographed for publicity purposes in this striking form before receiving its operating colour, in this case black. In March, it will be formerly allocated to Highbridge after a short period of trial running to iron out any faults. (Author)

have appeased the civil engineers somewhat when considering weight issues and the hammer blow effect on track and bridges.

A broad description of the S&D's new engines is easily dealt with. They made use of the G9 AS boilers and straight topped Belpaire fireboxes which had successfully equipped the Johnson 4-4-0 compounds when re-built under Deeley's management. These boilers, with Ramsbottom safety valves, produced a working pressure of 190lbs per sq in., were 11ft 11in long with an inside diameter of 4ft 6⅝in and contained tubes of two sizes – 148 x 1¾in and 21 x 5⅛in. With a firebox measuring 151sq ft the total heating surface was 1,321sq ft evaporative plus another 360sq ft for the superheater. For these engines, the team at Derby created their own version of Schmidt's smoke tube heater and this contained 21 flues set in three horizontal rows. The boiler was supported at the front end by a saddle casting and the smokebox was of a drum pattern. In this arrangement, the boiler was attached to the tube-plate in the same way as employed on the Deeley 990 Class 4-4-0s, which first appeared in 1907 and were then successfully modified with superheaters between 1910 and 1914. These, it seems, included improvements made by Fowler and Anderson, under patent number 12884/1911 and an improved automatic damper control, which the two men had also patented in 1911, under the number 2445.

These engines were fitted with two outside cylinders mounted high on the frames and set at an angle of 1 in 12 for clearance purposes. Walschaerts valve gear was employed as were 10in diameter outside admission piston valves with a lap of ⅞ inch. A third Fowler-Anderson patent was also included in the design, in this case one that provided an improved cylinder control by pass valve (patented three years earlier under the number 2446/1911), which was fitted to allow the engines to free-wheel downhill wherever and whenever possible. Oil to the cylinders and valves was provided by an 8-feed mechanical lubricator situated to the rear of the front coupled wheels. In addition, steam operated reversing gear was fitted, its ancillary equipment being mounted inside the main frames on the right-hand side of the engine.

Derby standard axle boxes, described as 'weak, undersized of the 4F 0-6-0 type', were used despite the fact that they were prone to running hot. However, they don't seem to have presented insuperable problems in service on the S&D and the position was improved with the provision of mechanical lubrication to the coupled boxes in a programme that lasted from 1919 to 1924. The coupled wheels had a diameter of 4ft 7½in and the pony truck 3ft 3½in. The coupled wheelbase measured 17ft 6in and if there were concerns that this might be too long for some of the S&D's curves, the flanged driving wheels and the pony truck, with its swinging links and radial arm, proved sufficient for the job at hand. To provide adequate braking on the line's steeper sections, Clayton is recorded as choosing to fit three brake cylinders to each engine. These were described as follows:

'The three rearward pairs of coupled wheels had brake blocks acting on their front faces operated by twin pull-rods inside the wheels from two 7½in brake cylinders under the rear drag-plates. The front pair of coupled wheels had brakes acting on their rear faces which operated a separate 7½ cylinder, which also operated the clasp brakes on the pony truck.'

When first installed, the brakes contained cast iron blocks, but these were found to wear too quickly and were replaced by a more robust Ferodo variety, which were deemed to be longer lasting, caused less brake dust and were easier to replace. Meanwhile, the need to increase wheel adhesion in slippery conditions, or when pulling heavy loads, was met by the addition of six sand boxes. These were fitted to the front coupled and main driving wheels to assist the engines when moving forward and on the back of the rear coupled wheels to aid reverse running.

The tender, although a Deeley design, was modified with a cab roof to offer the crew some protection from the elements when running backwards. These did not prove particularly successful or popular with the crew, though, and between 1918 and 1920 all were removed. Capacity-wise, the tenders could carry 3,500 gallons of water and 7 tons of coal. They were also fitted with two sets of Whitaker's tablet exchange mechanism – one each side to take account of the likely amount of reverse running necessary. Meanwhile, the cabs were laid out for right hand drive and in this state, they entered service,

By comparison with the other locomotives on the S&D, the new 2-8-0s must have seemed massive. Here crew and station staff happily pose with engine No. 80 at an unrecorded date but sometime before the tender cab roofs were removed post-Great War. When first entering service, these engines could not use the sheds at Bath because two bridges at their entrance were deemed too weak to take their weight. In due course, strengthening work was undertaken but until then, the 2-8-0s were based at Radstock. Here their height proved to be a problem by a matter of a few inches and the upper lip of the chimney was trimmed back, the dome cover flattened and the cab ventilator removed to allow access. (R. Hillier)

the last arriving at Highbridge in December 1914. Here they were greeted by a new Locomotive Superintendent, Mervyn Ryan having departed for Eastleigh.

Reginald Charles Archbutt was five years younger than Ryan, having been born in Derby on 8 September 1887, but they were near contemporaries whose lives followed a very similar educational and professional path to the S&D. He was the son of an Analytical Chemist and attended Derby School (St Helen's House) until 1902. He then went on to study engineering at the city's Technical School before being accepted for a place at University College Nottingham where he remained until 1907, under the tutorship of Professor Robinson. As with Ryan, his attendance was probably sponsored by the Midland Railway, which he joined in 1904, working at Derby during the summer vacations. On being awarded his Certificate of Association, he joined the works on a full-time basis and completed his apprenticeship in 1910, having spent a considerable period of time in the drawing office. Following this, he became an Inspector in the Locomotive Works, where he participated in testing, and also undertook firing duties as time allowed. He so impressed Fowler that the CME proposed his membership of the Institution of Mechanical Engineers in 1911, with Richard Deeley as supporter. So, in the parlance of the age, he was probably regarded as a 'coming man' and in 1913 this probably secured his appointment to Highbridge where he would remain until 1930.

With his background in inspection and testing, one of his tasks would have been to oversee the introduction of the 2-8-0s and assess the suitability of the design. This work began very shortly after engine No. 80's arrival in March. Without the refinements of a dynamometer car to aid this work, the tests carried out could only provide very limited information, but sufficient was gained to show the class to be powerful enough for the task for which they were built and fairly economical in their consumption of coal and water.

**With no** experience of the type to fall back on it was necessary to test the new 2-8-0s and see what they could do. Here, engine No. 80 begins a journey south from Bath in the spring of 1914 with a heavy down freight train, with engineers finding space to work and analyse performance in the indicator shelter around the smokebox. Here conditions would have been extremely unpleasant, especially when passing through the two long tunnels just south of the city. In the background, a banking locomotive, type unknown, can be seen assisting the ascent as No. 80 prepares to pick up a tablet and enter the single line section of track to Midford. (Author)

**Fowler 2-8-0** No. 83 is photographed in 1919 as it approached Cricklewood, apparently when under test by the Midland Railway for possible main line goods use. (DN)

However, the practicality of the design was also assessed, and some modifications were made in the years that followed. These included the removal of the tender cab roofs between 1918 and 1920, new brake blocks as described earlier, and then the removal of the steam operated sanding gear and its replacement with a hand operated gravity feed version. In addition, the superheater dampers were found to add little or nothing to the 2-8-0's performance and were removed.

There was nothing unusual in this. The need to review and

The S&DJR's role in war went far beyond the mammoth task of supporting the military machine as well as meet its other commitments to the community it served. There was also the effort and sacrifice of many of its employees in front line service, as this memorial in Highbridge bears silent witness. For such a small company, the number killed in action is quite remarkable even by the standards of this bloody conflict. This large bronze tablet was mounted on an external wall at Highbridge Station, having been unveiled on 8 March 1922 by Sir Allen Garrett Anderson. With the closure of the S&D, it found a new home in the town's Garden of Remembrance. It was designed by an S&D apprentice and made in the railway works, with all costs being met by staff from the locomotive, carriage and wagon department. (Author)

re-assess performance throughout the life of any machine then seek modification or replacement is second nature to a good engineer. And it was a cyclical process without end, as the history of the 2-8-0s would prove. Each CME in turn would tweak the design until it could be taken no further and obsolescence became a reality or, as in the case of steam power, the technology was ditched in favour of the internal combustion engine or electrification. But for the moment, the S&D had six powerful new engines and could enjoy the benefits that soon accrued from them. With the world descending into the chaos

**During the** early twentieth century, the S&D's safety record was considerably better than it had been in the decades before. This was largely due to the efforts of Dykes, Eyre, Whitaker and the civil engineers, but accidents still happened, as they always will even in the best run organisations. Just such a case occurred on Easter Monday 1914 at Burnham Station. In those days, local communities often arranged day out 'specials' to the seaside or somewhere else of interest. On this occasion, 0-4-4T No.52, when entering the station bunker first, jumped the points and was derailed taking the leading carriages with it. Luckily, any injuries were minor ones. The angle of the engine and first two coaches suggests that the train was probably moving a little too fast. The large crowd 'admiring' the driver's handiwork are all dressed in their Sunday best, indicating that they were probably the passengers. Meanwhile the driver or fireman poses for the picture from the cab, seemingly unaffected by the accident. (D. Neal)

of war, this was none too soon. By the time the last engine, No. 85, rolled from the works at Derby in late 1914, deadlock on the Western Front had become a reality and total war soon followed.

Its impact on the railways was immediate and all their efforts were soon subsumed by central government through its Railway Executive Committee, which had been set up in 1912 for just such a purpose. However, it only began to exercise real power on 4 August 1914 with general mobilisation and then controlled all aspects of railway work until 1921, under the chairmanship of Alexander Kaye Butterworth, who was General Manager of the North Eastern Railway. This committee, on which various senior managers from the industry sat, brought most independent action to an end, and placed strict limitations on new locomotive and rolling stock production. Everything had to be in support of the war and the engine policy reflected this strategy. If new locomotives were built, they had to be sanctioned centrally, which meant that goods engines were given priority for the duration. There were few, if any, exceptions, and the companies had to muddle through as best they

**The second** 2-8-0, No. 81, delivered to Highbridge was, according to the words written on the back of the print, photographed in 1918, though the location is not recorded. By this stage of the war, all the company's locomotives were suffering from reduced maintenance and the absence of cleaning staff, as the grimy condition of this engine confirms. It would be some time before the debilitating effects of wartime on the engines, rolling stock and infrastructure could be reversed, by which time the S&D had been absorbed by the LMS as part of grouping. (D. Neal)

could. Unavoidably, their engines and rolling stock were worked extremely hard and maintenance standards soon dropped as many trained men drifted away to the war. There was a gradual erosion of standards as the country and the railways wearied. This was only too apparent on the S&D, where its Traffic Superintendent, George Eyre, who had succeeded Robert Dykes in 1902, and Reginald Archbutt were increasingly hard pressed to keep the trains running efficiently, especially as demand increased. In this situation, the arrival of the 2-8-0s was a godsend.

Most railway workers were deemed to have reserved status at this time of national emergency, for the simple reason that no country could successfully fight a war if supplies could not be moved in vast quantities and on time. For some, the call to arms was too loud to resist for long and so the S&D lost many men, some never to return. Amongst their number would be the Locomotive Superintendent himself, who felt it impossible to remain at Highbridge for long. In June 1915, he sought James Anderson's permission to join up. Receiving an answer in the affirmative, he applied and was granted a commission with the Royal Field Artillery, with whom he served until 1919, seeing action on the Western Front and becoming a captain in the process. In his absence, Alfred Whitaker's son, Alfred Henry, became acting Superintendent, holding the post until Archbutt returned.

From the point of view of the locomotive fleet, the war years would prove to be extremely trying, but not a period of any great developments or changes. With traffic increasing to meet war needs and an expanding coal industry, the engines were hard pressed to meet demand as their condition deteriorated. The same can be said of all railway companies at the time, particularly the Midland, which for most of the war ran with James Anderson as acting CME, Henry Fowler's great organisational skills having made him an obvious candidate to join the burgeoning world of armaments production as Director of Production at the Ministry of Munitions. He later became Superintendent of the Royal Aircraft Factory at Farnborough, then added the titles of Assistant then Deputy Director-General of Aircraft Production to his task list. Throughout this trying period, he seems to have maintained a

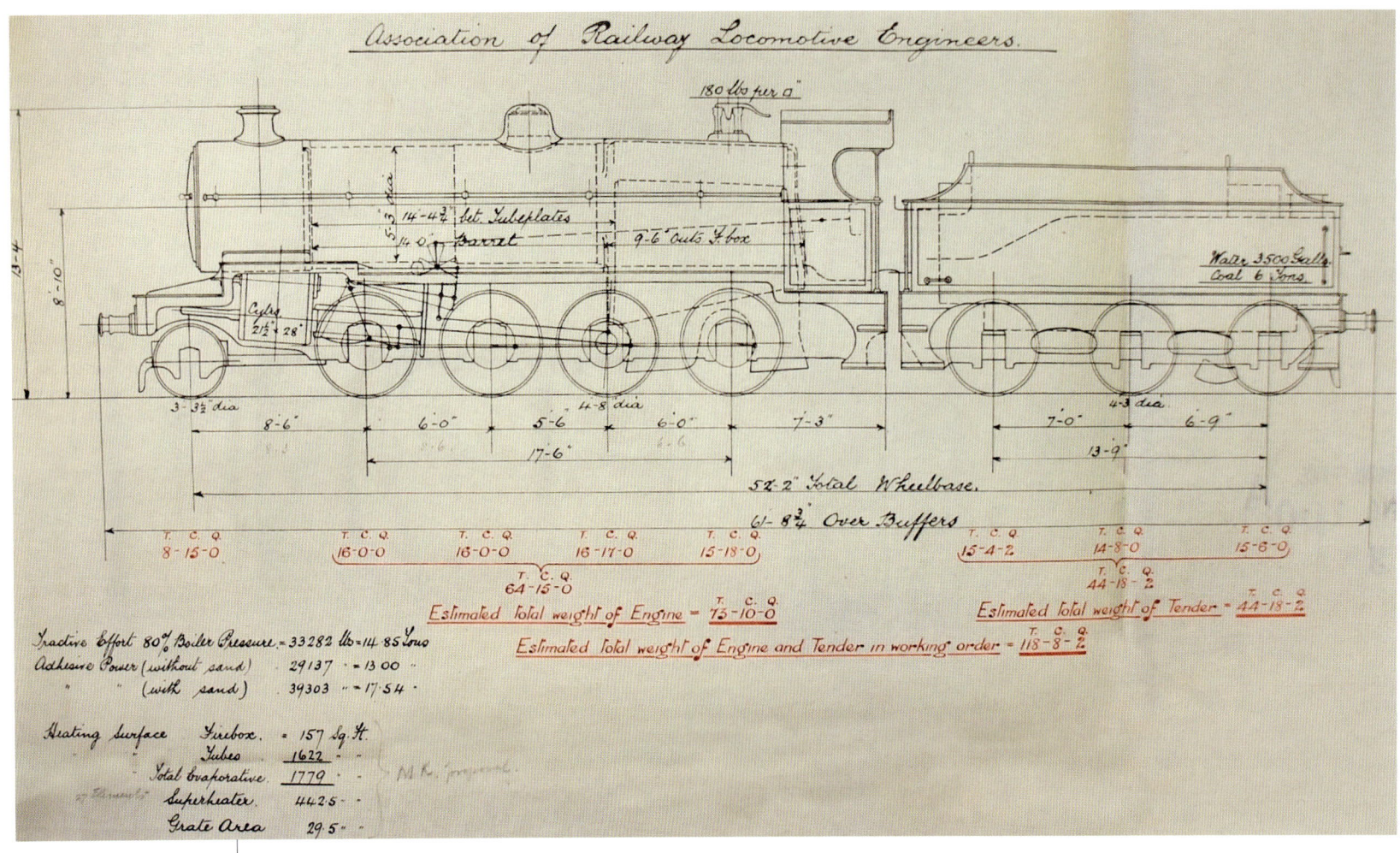

**The Fowler/Derby** wartime proposal for a new class of 2-8-0s containing many standard parts designed, as part of the Association of Locomotive Engineers wartime deliberations, to populate a post-war network that many thought should be nationalised. It wasn't to be, but this design does at least show how the S&D 2-8-0 might have evolved into a much larger, more widely used class. As it was, these ideas did help inform the second stage of the S&D's 2-8-0 programme which reached fruition in 1925. (R. Hillier)

watching, though often distant, brief on his old department. But under Anderson's leadership, the works ran effectively and gradually acquired many armament projects, including some linked to aircraft construction, to sit alongside its traditional tasks. And so it went on until the war ended and Fowler's return in May 1919.

There was some limited locomotive production at Derby during the war, most notably fifty 0-6-0 tender engines, later classified 4Fs, but little else. To fill in time, the reduced number of draughtsmen would have spent the war years considering future locomotive projects and any outstanding pre-war requirements. Undoubtedly they would have been encouraged in this by Anderson and also have been involved in producing standard designs for the centralised body, the Association of Railway Locomotive Engineers (ARLE), on which Fowler sat during 1917/18. Interestingly, one of their proposals built on the Midland's work in producing a 2-8-0 for the S&D but equipped with a number of parts that brought a degree of standardisation into play, possibly in expectation of railway nationalisation post-war. This didn't happen, of course, but the ARLE's work did provide a clear pointer to the future and the need for fewer designs containing interchangeable parts. In terms of its specification, the ARLE design went for Walschaerts valve gear, cylinders 21½ x 28in, plus coupled wheels ½in larger with slightly smaller wheels for the pony truck. It had a larger diameter boiler, in this case coupled to a Robinson superheater and 30sq ft of grate area and produced a total heating surface of more than 1,911sq ft with an expected 180lb. of working pressure, plus another 268sq ft for the superheater. By contrast, the S&D's engines had the Derby/Schmidt superheater and a total heating surface of 1,321 and 360sq ft respectively and a working pressure of 190lb. With these different combinations, the Highbridge locomotives produced 35,932lb. of tractive effort compared to a proposed 33,282lb of the ARLE proposal. To all intents and purposes they were twins, but with small variations that only a parent might know.

As this exercise passed into history and the railways sought to repair the ravages of war, their future was about to undergo radical change, though not under the nationalisation banner just yet. After much deliberation over the future structure of Britain's network, the government published its Railway Act during 1921 which created four new super companies under which most businesses would be grouped. Inevitably, there would have been a selection process and it might have made sense for the S&D to be entirely gathered in by the GWR. But on 1 July 1932, the S&D became jointly vested in the Midland Railway's descendant the LMS and the new Southern Railway, which had absorbed the LSWR, instead. So two long established relationships continued, for good or bad. Would the S&D have been better off with the GWR? Later history suggests that alignment under their banner might have seen the line starved of support and allowed to wither much earlier than it did in the late 1950s and early 1960s. So it may have been a case of better the devil you know. But when it came to the management of locomotive and rolling stock fleet, grouping did mean a possible change of direction, with George Hughes, late of the Lancashire and Yorkshire Railway, becoming the LMS's first CME with Fowler his deputy.

Grouping would come into effect in 1923, but before then the S&D's fleet would be subject to review as well as undergo a catch-up programme of maintenance. In fact, a wider review was commissioned late in 1921, led by Thomas Redhead, assistant General Superintendent of the MR. The terms of reference allowed him to consider all elements of performance in the face of diminishing returns and annual losses in excess of £70,000. The picture that emerged was a depressing one. Overstaffing was found to be widespread, in some cases quite excessively so, due, in part, to returning servicemen and the retention of their wartime replacements who would otherwise have been unemployed in a time of recession and few job opportunities. The control and movement of locomotives was also found to be ineffective and the workshops were criticised for poor and costly management of materials and maintenance work. In the light of this review, savings of £28,000 were identified, plus an additional £16,000 a year if Highbridge was closed and maintenance work transferred to Derby, with more minor repair work undertaken by running shed staff. With grouping, this controversial proposal was placed in the pending tray to await developments, while the other savings measures were implemented and, it seems, redundancies and transfers followed.

With losses on this scale and savings measures soon to be applied, one might have expected that any proposals to enhance the locomotive fleet would be held in abeyance. But this proved not to be the case. In April 1921, the exchange of three Johnson 4-4-0s for an equal number of rebuilt 483 Class engines went ahead with the replacements given the numbers of the superseded engines – 67, 68 and 69. Then, in early 1922, five 0-6-0 tender engines arrived from Armstrong Whitworth & Co. These had been ordered by Archbutt in July 1920, at an inflated cost of £10,960 each, before the parlous state of the S&D's accounts had been revealed by Redhead, which probably explains why the contract was honoured. It may have helped that the Midland had ordered another fifty of this type from Whitworth's and the government had introduced a compensation scheme that meant any difference between 1914 and 1918 prices was recoverable. Nevertheless, the transaction created an even bigger hole in company accounts. So one

**In 1922** a group of new 0-6-0s on their way from the Armstrong Whitworth works in Newcastle to Derby and then some to Highbridge. Three of these engines can clearly be identified as being destined for the S&D and the other two may also be in this group. This company, which was formed by Sir William Armstrong and Joseph Whitworth in 1897, by combining their already substantial businesses, went from strength to strength. They were already known for building ships, cranes, bridges, hydraulic machinery and much more, but together they soon expanded into automobiles, aviation and armaments. During the war, the company increased the size of their works to cope with many government contracts and then sought to fill the vacuum created by the Armistice with other work. This included a move into locomotive construction and the acquisition of contracts from the major railway companies. The fifty-five 0-6-0s for the Midland and S&D were part of this expansion and helped fill their Scotswood Works in Newcastle. In 1920 they had already built 200 2-8-0 engines for the Belgian State Railway and would, in due course, supply the LMS with 323 4-6-0 Black Fives, a class which would become very familiar on the S&D. In the post-war years, the company also recognised the benefits of diesel power and in 1919 moved into this field when acquiring the UK rights of the Swiss company Sulzer and its diesel engines. In due course, they began building diesel locomotives and railcars. (D. Neal)

assumes that demands on the locomotive fleet were deemed to be sufficiently pressing to warrant the acquisition of new engines. It may also have helped that when they arrived, four 1878 Neilson built 0-6-0s, Nos. 35, 36, 37 and 38, were withdrawn from traffic to provide some balance to the accounts. And to this could be added another four of the class which went in 1914 – two Neilson and two Vulcan built engines – but these reductions should more rightly have acted as a compensating saving against the six new 2-8-0s supplied that year.

Whilst the 2-8-0s were welcomed, they were still the product of a company that remained firmly wedded to a small engine policy. There was one other exception to this though. In 1919, the Midland Railway built a single 0-10-0 specifically to bank goods trains up

**Engine No. 59**, the third of the Armstrong Whitworth 0-6-0s, stands cold awaiting her next turn of duty. The other locomotive appears to be No. 58. In due course, these engines received the designation 4F, but when received by the S&D became 5P4Gs. They were painted black overall from new and so followed an instruction regarding colour schemes issued by the MR in 1921 for goods engines. In terms of dimensions, they were bigger than the last batch of 0-6-0s to arrive (in 1902 built by Neilsons). For example, they had slightly larger driving wheels at 5ft 3in, a 4ft 8in diameter boiler designated G7S, a grate area increased from 17½sq ft to 21sq ft, cylinders up from 18in x 26in to 20 x 26 and a total weight nearly 12 tons greater. They were also superheated and fitted with Belpaire fireboxes. Whilst these new engines were being built at Scotswood, some of the older 0-6-0s from 1890, 1896 and 1902 were beginning a rebuilding programme and fifteen would soon receive either G5 or G7 Belpaire boilers and new frames (a programme that began in 1920 and ran until 1928). In this condition, five of the 1890 built engines lasted into the 1930s, but four of the later engines stretched into the 1960s and others into the 1950s. The five Armstrong Whitworth engines lasted longest and were finally withdrawn from service as follows – two in 1962 and then one a year until 1965. (D. Neal)

the Lickey incline in Worcestershire. But other than this, any proposals for larger engines remained on the drawing board to await the arrival of their new LMS masters and, possibly, a change of direction.

As 1923 dawned, the S&D were far from being in a comfortable position operationally. The internal combustion engine was slowly but inexorably making inroads into their trade and they were having to absorb large cuts and, potentially, face even more cutbacks. Would the company have ready support within the LMS and SR, and would someone of influence step forward to become their advocate in securing the line's future? No one could guess what the future held or for that matter if the S&D would be allowed to wither on the vine, its role diminishing, in the decade ahead.

**By the** time grouping took place, the 2-8-0s had undergone several modifications, the most obvious to the cab, the tender having lost its roof. The engine pictured here (possibly No. 82 but this isn't entirely clear) is in modified form and in a suitably grimy condition to suit its hardworking role. (R. Hillier)

**This photo** captures only too well the nature of the S&D shortly before Thomas Redhead undertook his highly critical evaluation of the railway in 1922. A rural station, in this case Radstock, a beautifully turned out Johnson 4-4-0 (No. 18 with an H type round topped boiler it received in June 1911) and rake of carriages, but little sign of life in terms of passengers. However, this isn't surprising considering the area's low density population and the deep recession that afflicted Britain post-war. Goods traffic, linked to the mining industry, though still considerable in volume, was reducing as industry struggled to re-adjust to peacetime needs, and wasn't sufficient to sustain the railway. Grouping offered a lifeline but at a cost. (D. Neal)

# Chapter 5
# THE LMS YEARS (1923 to 1947)

When a new company is formed from many smaller ones, it is very easy for it to lose sight of its constituent parts. With the LMS, this was very much the case. It had a network reaching from London to Scotland and across into Wales covering nearing 8,000 miles of track, with 10,346 locomotives and employing some 230,000 people. It was a giant against which the S&D's small fleet of locomotives paled into insignificance. And it didn't help that the first few years of the LMS's existence would be deeply affected by the conflicting policies and working practices of the companies forced into this new alliance. The Midland Railway, the London and North Western and the Lancashire and Yorkshire were not ideal partners, having for too long ploughed their own furrow. And so the first decade was one of conflict as old allegiances stubbornly refused to disappear and some sought to dominate the new organisation and the way it worked.

This was a struggle that played out most graphically in the development of locomotives, with the small engine group,

**Five of** the men whose work would prove important to the S&D during the life of the LMS. Left to right – George Hughes the first CME (1923-25), Henry Fowler the second (1925-31) and finally the giant presence of William Stanier (1932-44, with Ernest Lemon filling in between Fowler and Stanier). Despite the skill of these three men, they needed talented and hardworking Chief Draughtsmen and the LMS possessed two of the most important to grace Britain's railway industry. Second from the right – Herbert Chambers who occupied the post between 1923 and 1935, then to his right the outstanding Tom Coleman with whom Stanier developed a relationship of huge importance and creativity. It was the product of their work that would provide the S&D with two locomotives of the highest quality in the years ahead – the 4-6-0 Black Fives and the 8F 2-8-0s. (Author)

led by James Anderson, now the LMS's Chief Motive Power Superintendent, determined to exercise their powers as they thought best. Anderson's ideas, apparently supported by the company chairman Charles Lawrence and his successor in 1924, Guy Granet, would hold sway for the most part, despite George Hughes and Fowler's best efforts to think more broadly. This is nowhere better illustrated than in the attempts both CMEs made in trying to bring a large Pacific into being to meet ever-growing loads, only to be thwarted by Anderson, or so it seems. Only with his forced retirement in 1931 and the arrival of Stanier a year later did this limiting policy slowly begin to unravel. Meanwhile, on the London and North Eastern Railway, Herbert Nigel Gresley had already seen the light and was rapidly taking his ideas forward with the first of his Pacifics and bigger freight locomotives, while on the GWR, Churchward was even more advanced in building engines of remarkable strength and reliability. On this basis alone, the S&D might have been better off if Swindon and not Derby had overseen their locomotive needs.

To provide continuity during these difficult years, Archbutt would remain Locomotive Superintendent at Highbridge, despite Thomas Redhead's criticisms of his department, and would stay so until posted back to Derby in 1930. During 1920, George Eyre had retired as Traffic Superintendent to be replaced by A.S. Redman, who himself made way for G.H. Wheeler in 1922, possibly a victim of Redhead's telling analysis. Wheeler, like Archbutt, would last until 1930 with the S&D. They still had an independence of sorts but little real ability to do more than manage what they were given, which meant struggling on with an established but generally underpowered fleet, with one exception – the 2-8-0s. However, despite the limitations in place, both Hughes and Fowler continued with an active development programme and pushed back boundaries wherever possible. For example, Hughes, who had little time as CME to make his mark before retirement, was able, with his Chief Draughtsmen at Horwich, John Billington, who died suddenly in 1925, and then Edward Gass until the end of 1926, to design a large two-cylinder 2-6-0 tender engine. Sadly, Hughes wasn't able to oversee the construction of the prototype in 1926 or the 244 that followed before production ended in 1932, but Tom Coleman, who followed Gass at Horwich, did, and then went on to produce bigger and better designs from Crewe, then Derby.

Operationally, the Horwich Moguls were generally well-received and were specifically noted for performing well when hauling heavy mineral trains over difficult terrain. As such they might have suited the S&D. In addition, Hughes and Billington also led in designing a 2-8-2 three-cylinder engine and a 4-6-2 four-cylinder Pacific operating with two sets of valve gear later modified to three-cylinders. Both of these classes reflected ideas being developed by Gresley at King's Cross, but unlike his engines, these types didn't find favour on the LMS and were almost dropped when Hughes departed. However, under Fowler both were briefly revived by the drawing office team at Horwich, but this time as compounds. Yet again the work was not allowed to continue, presumably because neither engine fitted into the Derby way of doing things, which in essence was the old Midland way as regards size.

Fowler's early years in charge contained many noteworthy developments, but, in essence, his output stuck reasonably closely to the prescribed small engine policy. However, he did seek to push back boundaries whenever possible. This resulted in the acquisition of 33 massive Garratt 2-6-0 + 0-6-2s between 1927 and 1930, the construction of 70 Royal Scot Class 4-6-0s in 1927 and 175 Class 7F 0-8-0s between 1929 and 1932. These locomotives probably demonstrated where his ambitions may have lain if unrestricted by the company's locomotive policy. However, his other work focused on types more likely to find favour, such as the Class 3F 0-6-0Ts, the continuing 4-4-0 programme with the arrival of the Class 2Ps in 1928, two classes of heavy tank engines and the 4F 0-6-0 tender engines. Then there were the 4P three-cylinder 4-4-0 compounds which first appeared under Hughes in 1924 but accelerated in number with Fowler in charge until there were 195 in service by 1932.

To help further these design goals, engine No. 67, one of the Derby built 483 Class of 4-4-0s which appeared in 1921, underwent dynamometer car tests in December 1924 and again in February and May the following year. These were part of a general review to compare the different

**With a** partner as large as the LMS, Archbutt must have looked at developments at Derby, Crewe and Horwich with great interest. If given free rein to pick engines to work the S&D, what might he have chosen from the many classes that Hughes and Fowler introduced? Papers held by the National Railway Museum suggest one intriguing possibility. In 1924, Fowler brought out the first of his Class 4 (later 4P) three cylinder 4-4-0s, which built on the success of the Johnson design, which was improved by Deeley and then modified by Fowler himself. In trials set up early in the life of the LMS, a compound, No. 1008, was matched to an LNWR Prince of Wales Class 4-6-0 and an ex-Midland 999 Class 4-4-0. No. 1008 was deemed to perform best. The tests were completed in 1924, but by then an order for twenty new superheated compounds had been approved with another 175 over the next eight years. The class became the subject of much publicity, which included the production of a brochure from which this picture is taken. In 1926, a single document held at York reveals that Anderson's Motive Power Department wanted to see some of these engines assigned to the S&D for 'more trials with a view to permanent allocation of five or six three-cylinder compounds of the latest type'. This doesn't appear to have taken place, presumably due to the high demand for these engines elsewhere. The reference to 'more trials' is a link to a short period in May 1925 when engine No. 1065 worked test trains from Bath to Bournemouth, the results being compared with a Class 2 4-4-0 over the same route. Would Archbutt have wanted these engines; who can tell? The chances are that he might, being a Midland man, but by 1926 he was probably more than happy just to have another five 2-8-0s. (Author)

**Although not** showing the Class 4P 4-4-0, No.1065, which ran dynamometer trials over S&D metals in 1925, this photo captures similar trials with her sister engine, No. 1094, a little later. Most documents describing this work no longer appear to exist and so it is left to the few memoirs of those who were there to describe the work that took place. (D. Neal)

classes of express engine the LMS had inherited from its constituent companies. Ernest Cox, who was at that time based at Horwich, recalled that these trials were really set up to 'establish the ascendency of the Midland Compounds'. To do this, various compound engines were assembled over a two-year period and then run over routes between Leeds and Carlisle, Preston and Carlisle and Bath to Bournemouth. The engines included were No. 388 'a Prince Class 4-6-0' and No. 2221 'a Claughton 4-6-0' both from the LNWR, two Caledonian 4-4-0s, a Midland two-cylinder Class 4, an S&D 4-4-0 and a Horwich built four-cylinder 4-6-0, No.10460, which had been rebuilt with a superheater, Walschaerts valve gear and piston valves in 1919. Cox then described the nature of this work and the results:

'All tests were repeated with 300 and 350 ton loads except the Caledonian engines which only took 300 tons, while the Claughton and the Horwich rebuild took 350 and 400 ton trains. The tests on the Bournemouth line were loaded to 220 tons.

'As against coal consumption of the order of 5lbs. per Draw Bar Horse Power, which was general with simple expansion on the LMS then, the Compound ranged around 4lbs on comparable duties and it was little wonder that it was accepted as being head and shoulders above its contemporaries in this respect.'

Despite this Cox noted that:

'amongst all the self-congratulation that followed these trials was sounded a report by Collett, CME at Swindon, to the effect that one of the new Castle Class locomotives on the GWR had run a test train at a coal rate of 2.83lbs per D.B.H.P. As we could think of no earthly reason as to how this could have been achieved we simply preferred not to believe it!'

Stanier's arrival in 1932 from Swindon would throw some light on this issue to the benefit of the LMS but also the S&D.

The most notable omission in this programme was a 2-8-0 type which had clearly interested Fowler when building six for Highbridge in 1914, and then pursued when developing a standard range of locomotives during the war as part of the ARLE's deliberations. With the success of the S&D engines, one

**The details** provided with this print suggest that this photograph was taken in 1920 near Wincanton shortly after engine No.83 had had its tender cab roof removed. Confirmation of this date is difficult as is the summary that it was taken about the time the engine was loaned to the Midland Railway for testing purposes. Fowler and team were, apparently, developing a new version, building on the CME's work with the ARLE, and these trials allowed them to study the 2-8-0 at closer quarters. However, their design work got no further than the drawing board. (R. Hillier)

would have thought that the type would have found ready acceptance within the Midland and then the LMS. In fact, at least two of these engines, Nos. 83 and 85, were loaned to the MR for evaluation in 1918 and again the following year. Though found capable of pulling heavy loads with comparative ease, the nature of the task involved longer distances and higher speeds than those achieved on the S&D, which brought to light a problem with their axleboxes. It seems that this was corrected at Highbridge by extending mechanical lubrication to the boxes sufficient for local needs, but this was insufficient for the Midland Railway and problems arose. Despite this, one source suggests that a proposal to build a new 2-8-0 class based on the S&D's design was prepared at Derby in July 1920 and submitted for approval. However, it seems that its likely axle loading of 17½ tons proved too much for the civil engineering department. As a result, the project slipped into the pending tray until Ernest Lemon became CME for a short period in 1931 when the idea was revived without success. It was then left to Stanier, who took over in 1932, to take the concept to a successful conclusion much later with his Class 8F 2-8-0s.

Records in public and private hands are not completely clear when it comes to confirming how the 2-8-0 programme evolved. It was left to Ernest Cox to provide some commentary and clarify the issue. He was well placed to do so. He moved from Horwich to Derby in 1925/26 where he remained until 1930 and during this period was involved in locomotive testing. After this, in early 1931, he was appointed to Euston where he worked for the CME's Locomotive Assistant, Sandham Symes. In both these posts he was well placed to observe how ideas were developed, selected or rejected. When remembering the debate over the proposed 2-8-0s he recalled in a letter written in 1962 that:

'Lemon began considering future heavy freight engine requirements for the LMS during 1930. There was some data available from trials ten or so years earlier with the S&D 2-8-0s and some rough drawings of a proposed version for the Midland Railway which had been rejected, to help him. He then instructed us to undertake a detailed investigation of the engines then available to see what if anything their designs could tell us. These included ex-LNW G1 and G2 Classes (both 0-8-0 types), the new Class 7 0-8-0s and finally the Somerset and Dorset 2-8-0s. Although few in number they were the quintessence of Derby practice according to the Fowler/Anderson school of thinking. Engine trials followed and one of these entailed a 7F 0-8-0 being detached, with dynamometer car, to Bath to be compared to an S&D engine. Although interesting these tests told us little we didn't know already. The North Western types were the best of their kind but nobody wanted to build more of this old design, although they proved to be the most thermally efficient. The standard 7s had proved to be a great disappointment and generated a high number of hot boxes and one broken crank axle in service making them unreliable. At the time the word used to describe them was "undershod".

'When it came to the S&D engines our assessment was also helped by the fact that three of them [ Nos. 9671, ex-81 as renumbered by the LMS, 9676, ex-86, and 9680, ex-90], had been loaned to the Midland Division to help out when their standard 7s fell in numbers due to failures in service. However, the S&D engines were also prone to hot boxes at times when run over long distances, but they were in all other respects mechanically sound as proven by the documented trials in the immediate post-war period and analysis of their subsequent performance. However, its cylinders proved to be too wide to permit better route availability. Thus none of the existing types to which the LMS had access provided sufficient scope for further development and so three alternative schemes were prepared – a 2-8-0 development of the Horwich 2-6-0s, a 2-8-0 version of the 7F 0-8-0 with outside cylinders and a massive 4-8-0 with an enlarged Claughton boiler. The CME was unable to choose between these alternatives and simply ordered fifteen more of the unsatisfactory 7F 0-8-0s, with modifications. It was left to Stanier and Coleman to find a better solution later in the 1930s.'

Although most of Hughes and Fowler's work as CME had minimal impact on the S&D, some did and in due course a few of their engines would find their way onto the line – the 4Fs, the 3Fs and the 2Ps. But before then, the volume of traffic, the deteriorating condition of some of its older 0-6-0 engines and the continuing need to economise by reducing double-heading made it necessary to seek replacement locomotives. The 2-8-0s had proved essential in pulling heavy mineral trains with minimal assistance and in so doing kept services to schedule. However, it had been shown that at any one time, two or even three might be undergoing maintenance or repair leaving too small a number to meet the company's freight commitments effectively.

Although the S&D's finances were still proving to be something of a challenge, when faced with a sound spend to save proposal, few accountants would be unlikely to stand in the way. In this case, it helped that the 1921 Railway Act had introduced a compensation scheme, mentioned in the last chapter, which created a pool of funds that might be used to cover some of the cost of replacing engines worn down by war service. With the S&D offering up five

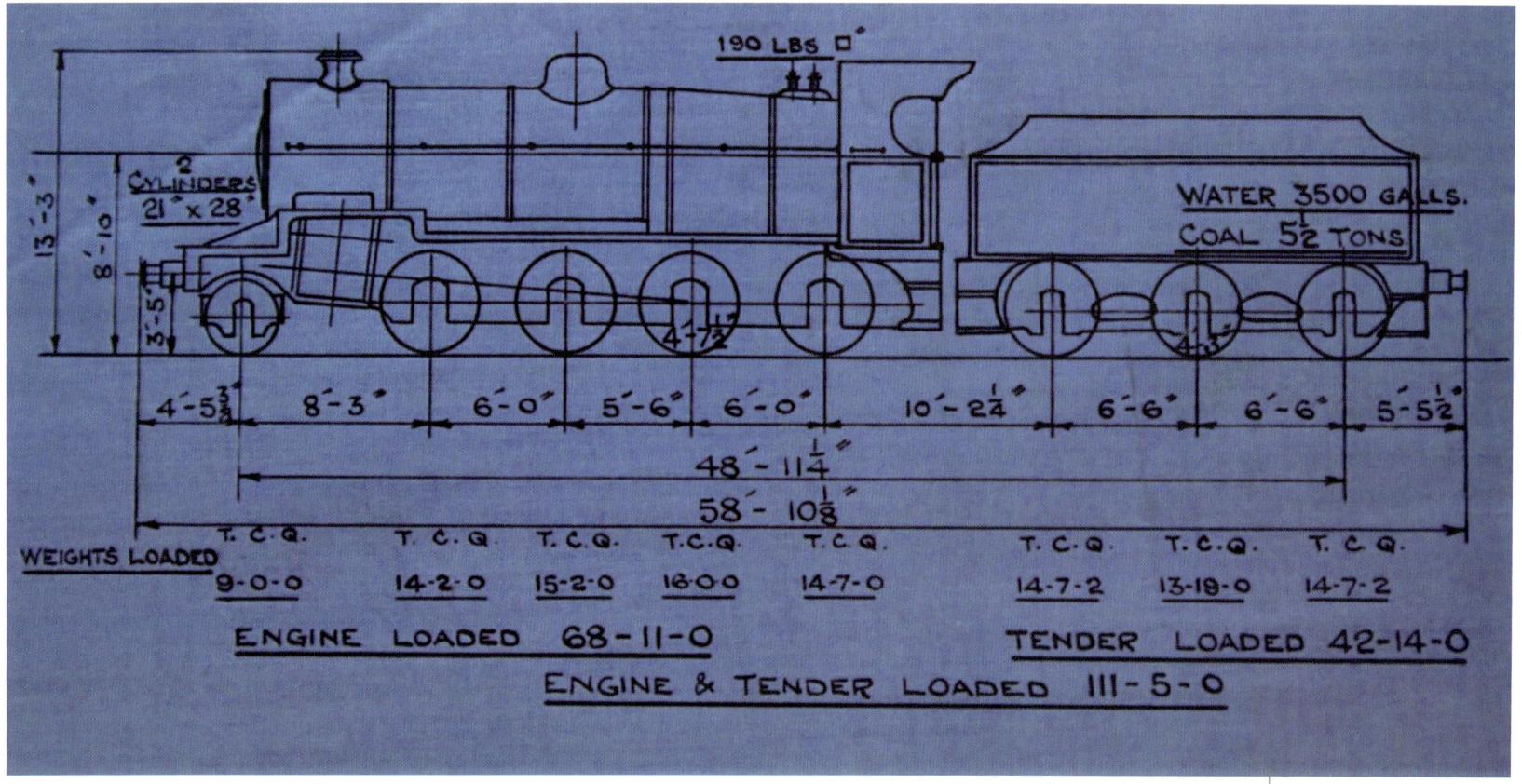

**A very** early drawing prepared in 1924 to show some of their dimensions of the five new 2-8-0s. It was part of the overall specification document sent to Robert Stephenson & Co to assist the bidding process. (D. Neal)

old and tired 0-6-0 tender engines as a compensating saving, plus government funding, the cost of new locomotives could be met. As a result, Hughes approved an order for five additional 2-8-0s in July 1924, with the work contracted out to Robert Stephenson & Co Ltd at a cost of £6,570 per engine, significantly more expensive than the first six locomotives built in 1914 at Derby due, in part, to spiralling inflation. Why this task was outsourced by Derby isn't clear, but it may be assumed that the LMS's workshops may have been full to capacity with other work. However, Stephenson's were locomotive builders with a very long history and a very good pedigree, which was enhanced during 1902 when the directors chose to centralise the business in new works at Darlington having been based in Newcastle since the 1820s. By 1924, the LSWR trained Clarence Goodall was Managing Director, having earlier been Works Manager. During his time in charge, the company had built many locomotives for customers in Britain and overseas, most recently eighty-two 2-8-0s for the War Office and thirty-five Class 4300 2-6-0s for the GWR. By 1924/25 they appear to have been struggling for large orders and the five S&D engines would have provided a reasonable fill in task until something more substantial turned up.

In terms of design, the new engines differed to the earlier 2-8-0s in a number of ways, but there were two very obvious changes – the type of boiler fitted and a smaller tender. The first of these modifications was an interesting one, because the G9AS type boilers used on the original six engines were thought to have been more than adequate for the work they did. But when designing the final five the drawing office team led by Herbert Chambers decided to fit a bigger, non-standard type boiler with the designation G9BS. Over the years, there has been some speculation by historians who were trying to understand why this change took place,

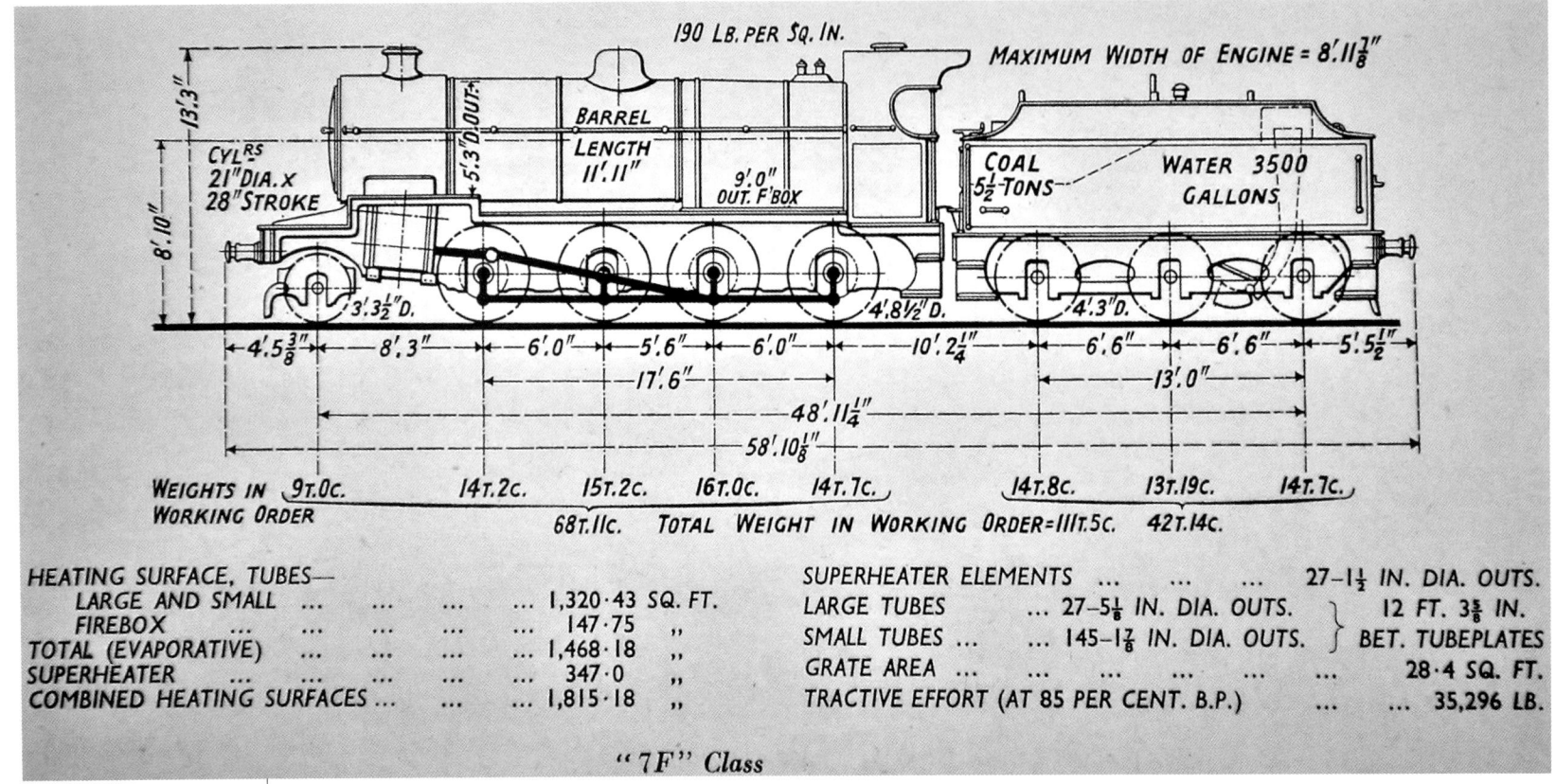

"7F" Class

The slightly more refined diagram produced by drawing office staff in 1925 to provide managers and workers with a simple guide to the class. (Author)

apparently without resolution. The most likely explanation is that Chambers and his team were eager to enhance the performance of these engines and felt that the G9BS would achieve this, with the added bonus that the overall increase of 2 tons and 6cwt between the types would increase these engines' tractive effort to adhesive weight ratio from 3.5 to 3.7. In doing this, they could have been responding to a request from Archbutt for more power, in which case this was a simple way of achieving this without a radical redesign of the type. It was work that also fed into Fowler's continuing search for a new freight locomotive for the LMS, which by 1924/25 had included a new 2-8-0 and was coming round to a 2-8-2 as a possible solution, mirroring Gresley's work on the LNER with his P1 Class.

In 1926 this resulted in outline plans for such a locomotive being produced which was, as Ernest Cox later described it:

'A scaled up version of the latest S&D engines fitted with a trailing truck, but in this case a compound locomotive with four-cylinders weighing 151 tons plus. However, like his earlier attempts to get away from the Midland's small engine policy it failed, taking with it his attempt to revive Hughes' work and build a Pacific Class engine as well'.

The new boiler had increased evaporating and superheating surfaces and a different number and size of tubes. The G9AS boilers were built with 148 x 1¾in and 21 x 5⅛in tubes, the new G9BS with 145 x 1⅞in and 27 x 5⅛in. The superheater itself was again a Derby version of a Schmidt design but in this case, there were twenty-seven flues in three horizontal rows rather than twenty-one. The grate area remained the same, as did the working pressure of 190lb, but whereas the earlier engines had been fitted with Ramsbottom safety valves, the new locomotives received the more recently developed R.L. Ross & Co 'Pop' version, in which William Hargreaves, the managing

**Part of** a Robert Stephenson & Co publicity brochure from the early part of the twentieth century. Why the LMS decided to sub-contract the construction of five new 2-8-0s for the S&D to this company isn't entirely clear, especially as the workshops at Derby manufactured many of the parts. It may simply have been a case of there being too heavy a workload in the works at Derby, Crewe and Horwich at the time. These engines were the last built by this company for the LMS in a relationship that stretched back to the earliest days of the Midland Railway. Locomotives would continue to be built at their Darlington Works until 1960, by which time Robert Stephenson and Co were part of English Electric. (Author)

director, played a leading role. With a boiler having a diameter of 5ft 3in, seven inches greater than the 1914 batch, it was necessary to keep the vertical clearance in check by redesigning the smokebox saddle and boiler fittings and fitting a shorter chimney and dome.

Unlike the earlier engines, the second group were left hand drive which meant some re-adjustment to the cab layout. For example, the vacuum ejector and hand operated reversing gear, which on the other engines had been steam operated, were moved to the driver's side. There was no repeat of the tender having a section of roof, instead the cab's roof was extended back a little to try to improve life on the footplate for the crew. In addition, all the other modifications made in service to the engines with smaller boilers were replicated here. However, the steam operated sanding apparatus which had been fitted to engines 80 to 85 was replaced by an Henri Lambert designed system which appeared in the years before the Great War in France. His invention applied a mixture of sand and water to the rails. By doing this, it was found that the sand could not be blown off the track by strong cross winds and it removed the need to keep the sand in a dry state before the apparatus was filled. There was also the added benefit that it could be set up to release the mixture in a continuous way. This, it was claimed, would be advantageous on long steep gradients and when passing through the long tunnels immediately to the south of Bath where water ingress made the rails greasy. It was believed that

**All five** large boilered engines, given the numbers 86 to 90, were built in 1925 and, after a period of trial running at Derby, reached the S&D that year. As was generally the case each engine was photographed in primer grey before its overall black scheme was applied. Engine No. 90 is captured here in its interim state. (Author)

**Before being** allocated to the S&D, engine No. 86 was put on display at the July 1925 Railway Centenary Exhibition at Darlington to commemorate the opening of the Stockton-Darlington line. The event was attended by the Duke and Duchess of York and part hosted by Nigel Gresley. In the upper picture, he can be seen showing the Duchess some of the exhibits while, to their right, the S&D 2-8-0 can be seen (second locomotive from the front). (Below) A clearer view of No. 86 showing a model of *Locomotion No. 1* displayed on its running plate. The two engines together like this presented an interesting juxtaposition between the latest Stephenson built locomotive and the famous 1825 creation of George and Robert Stephenson. (R. Hillier)

the Lambert system would also allow an engine so equipped to pull a load larger by at least 10 per cent greater than comparable engines with steam operated gear. The manufacturer's brochures made great play of this and quoted usage figures achieved by the Paris-Orleans Railway as evidence. How far this was proven when in service with the S&D does not appear to have been recorded or analysed.

The tenders fitted to these new engines are described as a standard high sided Fowler variety. Their wheelbase was exactly 13ft and they were 9in shorter than those attached to the first six engines. Whilst the water capacity was the same at 3,500 gallons, they carried 5½ tons of coal rather than the 7 tons of their sisters. In case the engines should be needed in future to work on the LMS's main line, where water troughs were a part of daily life, the 2-8-0s were fitted with some, but not all of the pick-up equipment necessary to do this. There seems to be no evidence to suggest that these engines were ever fully fitted, but they may have been when they were later involved in trials or were loaned to the LMS for short periods.

So the new engines entered service and, like their sister engines, appear to have been well-liked by the men who crewed them. However, both types appear to have gained a reputation for being erratic steamers at times for reasons that are not entirely clear. By trial and error, various experiments with different size blast pipes gradually helped achieve a better balance and the problem appears to diminish. But if a driver or fireman was ham-fisted, or inexperienced on the class, the problem re-occurred no matter how good the engine. Equally, the quality of the coal used was important. This was especially so during the two world wars, when the quality was often poor, reports from across the network suggesting there was often more dust than lumps in tenders as the mines struggled to produce sufficient coal for all Britain's needs.

Meanwhile, the perennial overheating problem associated with their 4F 0-6-0 derived axle boxes, highlighted when tested by LMS crew as part of comparability trials, though apparent when operating over the S&D, does not seem to have been of major

**In 1930,** as a measure of economy, all the S&D's locomotives were taken into the LMS's stock when the Locomotive and Carriage Department at Highbridge was wound up and their work transferred to Derby. The engines were soon given LMS numbers. For example, the 2-8-0s changed from 80 through to 90 to become 9670 to 9680. In this new guise, No. 86/9676 is seen working a heavy goods train at an unrecorded location. These numbers lasted until 1932 when the engines became 13800 to 13810. (Author)

concern. This was undoubtedly due to the shorter distances worked, lower average speeds and comparatively low average yearly mileage of 56,000 miles. However, their coal consumption at 87.4lb. per mile was considered heavy, especially for a company being stretched financially, though there were compensating savings due to reduced double-heading and banking. Interestingly, though, when worked by LMS crew during trials, the engines produced marginally lower consumption figures of 80.6lb. per mile.

As experience of these engines grew, some of the changes made in their design were rolled out over the earlier members of the class. However, they were fairly minor in nature. For example, these included the removal of Ramsbottom safety valves and their replacement with the Ross Pop versions. The steam operated reversing gear gave way to the hand-operated screw type and the cab roofs were extended back towards the tender. Nevertheless, all the modification work was not all one way. In due course, it was decided that, as a measure of standardisation, the G9BS boilers be replaced with G9ASs when general repairs fell due. One wonders if this action could also have been in response to concerns expressed about the G9BS's comparable performance and economy? I could find no evidence either way, so it remains a possibility only. The first engine, No. 89, was modified in 1929/30, soon followed by No. 90 in March 1930. But the others did not follow suit for some time because a small pool of spare boilers was available to keep them going as they were.

The safety record of the 2-8-0s was fairly good but at times accidents simply happened. Just such a tragedy overtook engine No. 89 at 6.35 pm on 20 November 1929 when pulling a coal train

**The scene** of devastation that greeted rescuers during the evening of 20 November just beyond the entrance of the engine sheds in Bath. Engine No. 89 had come through Combe Down tunnel out of control when hauling a heavy coal train from Radstock. It then came off the track, rolled over and crashed through some wooden buildings where it came to rest, as witnessed by this picture taken a short time after the accident. (R. Hillier)

**In the** early hours of 21 November, the full extent of the damage caused by engine No. 89 was only too apparent. By this stage, the bodies of the three dead men had been removed and a heavy crane was beginning to lift the wagons away so that access could be gained to the locomotive. This also allowed Colonel Trench, the Ministry of Transport's accident investigator, examine the engine in greater detail. (Author)

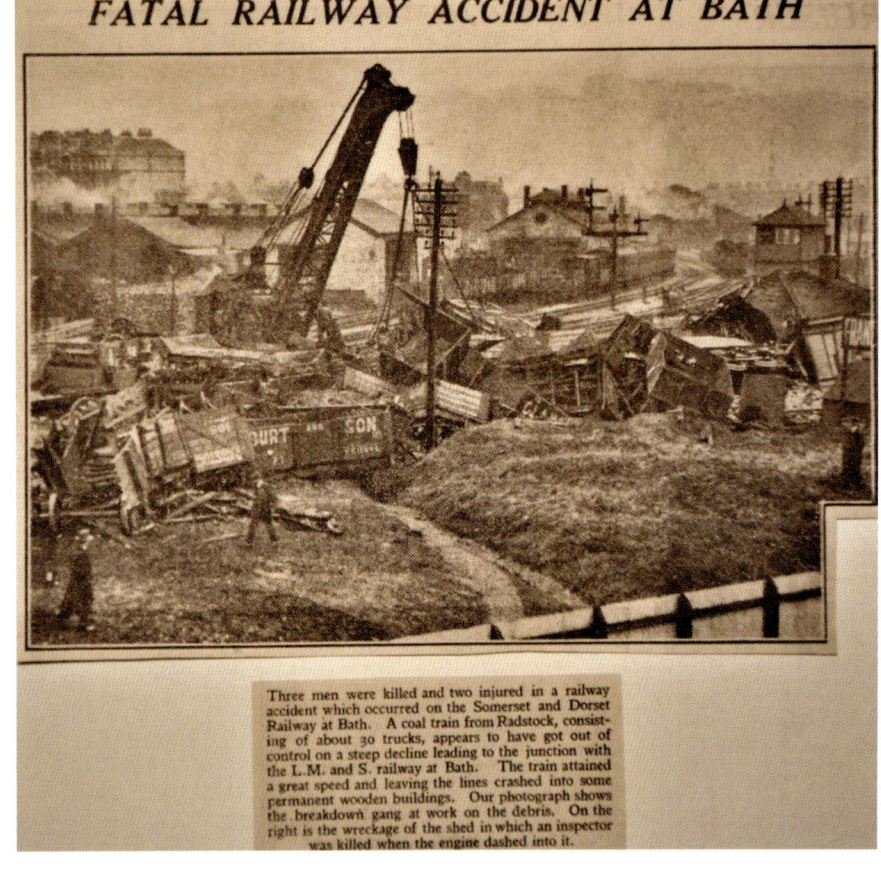

Three men were killed and two injured in a railway accident which occurred on the Somerset and Dorset Railway at Bath. A coal train from Radstock, consisting of about 30 trucks, appears to have got out of control on a steep decline leading to the junction with the L.M. and S. railway at Bath. The train attained a great speed and leaving the lines crashed into some permanent wooden buildings. Our photograph shows the breakdown gang at work on the debris. On the right is the wreckage of the shed in which an inspector was killed when the engine dashed into it.

from Radstock tender first. Sadly, this resulted in three deaths. As with any serious accident, the temptation to jump to conclusions is strong. The need to find a cause, apportion blame or simply change a dangerous practice or correct a fault in equipment is very pressing, but due process has to be followed and any initial conclusions have to be subordinated to an official independent investigation in this case by the Ministry of Transport. Their representative, Colonel Arthur Trench, late of the Royal Engineers and an investigator of considerable experience, was soon on the scene, though in the dark there was little he could see or do. For the first few desperate hours, the rescuers' first duty was to help the wounded then remove the dead. Yet even at this stage, some important pieces of information might be gleaned to point the way towards a likely cause. This came when the driver, J.H. Jennings, and fireman, Pearce, were discovered one on top of the other, both injured. For some reason, Pearce had wrapped a coat around his head, and it was established that they had been lying on the cab floor when the accident occurred. When able to get into the cab, Trench observed that the regulator was still open, the reversing wheel was nearly in full back gear for tender first travel, the ashpan dampers fully open and the tender brakes in full off position. All this pointed towards an uncontrolled descent through Combe Down tunnel before the crash and the possible immobilisation of the crew for some reason.

By degrees Trench slowly put together the facts of the case and concluded that:

> 'There was no question of defective condition or operation of points, signals or permanent way. There is no doubt that the speed of the train at the moment of passing Bath Junction was much in excess of what was normal or safe. The derailment was the natural and almost inevitable result of excessive speed.'

His 28 February 1930 report then debated the likely reasons for this and reviewed any supporting evidence. In doing this he quickly eliminated 'careless or reckless driving on the part of the driver' and 'the overpowering of the engine by the weight of the train and/or brake failure'. This left the only reasonable alternative being 'that during the descent from Combe Down Tunnel neither of the men on the footplate were in a condition capable of taking any action to control the train'. The cause of this was believed to be the 'hot and smoky' conditions in the tunnel and the engine's 'unusually long time in the tunnel and the inference that the engine was labouring heavily for most of the time', so causing even heavier volumes of smoke and fumes to accumulate in the cab. Sadly, Jennings died shortly after rescue and before he could be interviewed, but Pearce, probably

**How the** second large boilered 2-8-0 appeared after being fitted with a G9AS boiler in March 1930. During the year, the LMS's numbering system had been extended to cover S&D engines and No. 90 had become 9680. (R. Hillier)

because he had covered his nose and mouth, survived to bear witness to these events, although seriously injured. In this he was supported, as far as possible in the circumstances, by the train's guard, Wagner, who had leapt from his van shortly before the impact and survived, albeit badly injured. Their evidence allowed Trench to conclude that:

'Both driver and fireman were overcome by smoke and fumes while passing through Combe Down Tunnel, and that the engine emerged uncontrolled. It can be shown by calculation that it might well have attained a speed of 50mph at the foot of the gradient in spite of the guard's brake application… Post-mortem examination of the driver found there was at least 75 per cent saturation of the blood with carbon monoxide and he must have been unconscious and helpless to perform his duties.'

Although he was cleared of any responsibility for the accident, Jennings's widow Ellen could take little comfort from this or, for that matter, could the families of the other two men killed. They were 'Inspector [Isaac] Norman who was on duty in the yard and [Sydney] Loder, a clerk in railway employ who had just arrived at Bath Station and was making his way home, who were caught by the debris and killed'.

Within days, the wreck had been cleared, the engine being found to be in reasonably good condition in the circumstances. Repairs were soon carried out and the opportunity was taken to remove No. 89's G9BS boiler and replace it with a G9AS version. In this state, the engine returned to service early in the New Year, a packing piece joined to the smokebox saddle to carry the smaller smokebox.

Events such as this have to be endured and for a time the spectre of this accident would have cast a heavy shadow over day to day operations of the line. This would have been particularly so for each crew passing through the often-toxic atmosphere of the Devonshire and Combe Down tunnels. To provide some re-assurance and help identify ways of alleviating the problem, the company conducted a number of trials to establish the extent of the risk involved. There was, of course, much experience of the tunnels already over fifty or so years of operation, and few if any drivers or firemen had thought the risk excessive during those years or were too frightened to speak out and risk disciplinary action or unfavourable treatment. In the few accounts of life on the line that exist, it is clear that crew found the passage through the tunnels trying at times, especially if the smoke from a preceding train had not dispersed fully and the weather was damp and humid. But, in general, it was accepted as an inevitable part of life that had to be endured with the help of cloths or handkerchiefs held over mouths and noses. Health and Safety legislation was then in its infancy and was often honoured more in the breach than the observance by employers, so standards of safety were often set very low. However, the trials revealed very little, except that footplate crew tolerated the discomfort with great stoicism.

So life went on and the passage of time and normal operations soon calmed frayed nerves. The locomotives still underwent changes, the reboilering and modification to existing engines went on and Archbutt continued searching for ways of enhancing his fleet. Here Derby's development work offered some fresh options to be explored, the first of these being the Fowler Class 3 (later 3F) 0-6-0 tank engines that first made their appearance in 1924. By 1928/29 they were well-established, and the S&D had acquired seven of the 422 built by contractors or the LMS over an eight-year period ending in 1932.

In a report submitted in March 1928, presumably with Archbutt and George Wheeler's input, the extreme age and poor condition of fourteen engines was highlighted – two passenger and twelve goods locos. It was made plain that rising maintenance costs, including the requirement to replace boilers, frames and cylinders, could not be justified. So a case was made for their replacement. The analysis of cost, presumably aided by staff at Derby, made this a clear spend to save programme, especially when other more drastic, but long planned changes were added. In this form, the plan found ready acceptance. Seven Class 3Fs at £3,000 each and three superheated 4-4-0 tender engines, costing £5,000 each, would be purchased to replace these obsolete locomotives. It was estimated that the reduction of five of the oldest engines would save £8,750 per year in maintenance costs alone. But here a political twist was added to cover the remainder of the programme's costs. By combining the LMS and S&D's motive power establishments at Bath, a further saving of £1,086 per annum could be achieved, but this would be dwarfed by another £29,000 per year if the works at Highbridge closed too. However, while adequate repair facilities could be found at Bristol and Derby to service the fleet, the closure of Highbridge was probably deemed too sensitive an issue, particularly in the face of the worldwide economic slump that descended at that time. So Highbridge remained open, for the time being anyway, and the curse of redundancies was avoided for another few years. On this basis the report was accepted, and the engines ordered.

The Class 3 0-6-0Ts were descended from Johnson's Midland Railway 2441 Class tank engines that first appeared in 1899, with sixty being built over the next three years. Under Fowler, they underwent modification which resulted in Belpaire fireboxes being fitted to G5½ boilers and the cabs being redesigned. Production commenced in 1924 and would continue on until 1931 by which time 422 had been delivered, all but 15 built by contractors. The S&D's seven engines were constructed by W.G. Bagnall Ltd of Stafford, who in total built thirty-two of the class, the first in 1926, the last being those destined for Highbridge.

While this was happening, the future of Radstock's three shunting engines came up for consideration and it was decided to replace them with two modern 200hp doubled engined 0-4-0T locomotives at a cost of £1,950 each. These coal fired engines were built by Sentinel at their Wagon Works in Shrewsbury, to where the company had moved

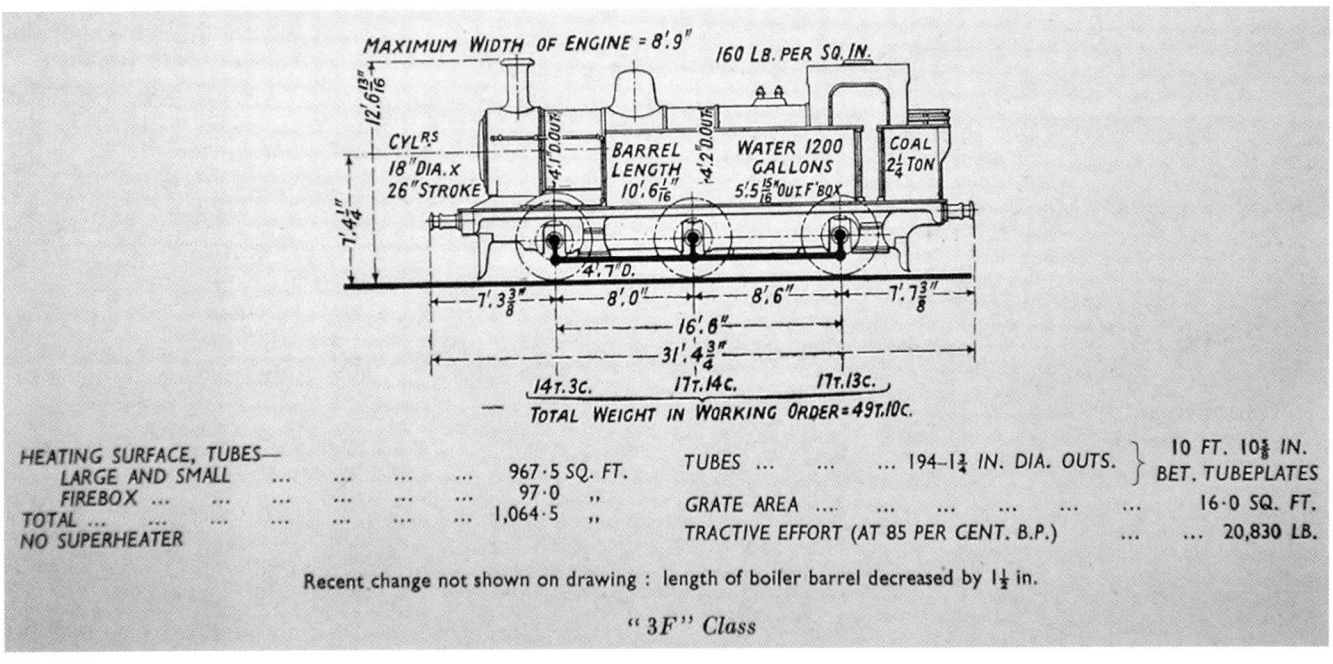

"3F" Class

(Top) The basic dimensions of the LMS Class 3F 0-6-0Ts which began to appear in 1924 but didn't reach the S&D until December 1928/January 1929. In addition to the details given in this diagram, these engines had two cylinders measuring 18in x 26in and a 5ft 6in firebox. The seven locomotives supplied to the S&D were given the numbers 19 to 25. These then became 7150-7156 in 1930, and finally 7310-7316 in 1934 in which state they remained until the railways were nationalised in 1948. When built, these engines were all painted blue, probably in recognition of the mixed duties they would perform – pulling blue carriages on local passenger services as well as banking, shunting and main line short distance freight duties. It is in this condition that two of the engines, Nos. 24 and 25, are pictured above. In due course, they would be painted black with the last remaining in service until 1967. (Author)

in 1915. With a history of producing steam powered lorries and buses, it was a natural step for them to move into the locomotive business and by 1928 they had developed a range of engines and railcars. By then, the LNER had become one of their biggest customers, having ordered twenty-four Class Y1 0-4-0Ts which appeared between 1925 and 1927, plus thirty-two Y2s, and were well into a programme for eighty steam railcars. Sentinels were also servicing overseas contracts as well as its domestic programme, which would, in due course, include the LMS, as well as the S&D. However, as petrol and diesel engines became more common, the company gradually adopted this technology when producing road vehicles from the mid-1920s. Nevertheless, steam continued to dominate their locomotive programme with the last two examples being built in 1958. When these engines entered service a brief description of them was circulated. It read:

**The casual** observer might be forgiven for thinking that the S&D's two Sentinel engines were diesel powered, such was their shape. That is until they began hissing and emanating steam when their source of power belied their lines. When supplied, the engines were numbered 101 and 102 but became 7190 and 7191 post-1930. These two photographs above capture both engines in their S&D and LMS states.

'They are very powerful examples of their type weighing no less than 27 tons in working order. They have a relatively short wheel-base of 5ft 6in. with outside wheel bearings and the axles retained in correct relation to the driving chains by radius-rods connected to the axleboxes. They are driven by a pair of 100 hp high speed vertically enclosed engines, each with two cylinders each measuring 6¾in by 9in., driving on to one reduction gear below them. From this the power is conducted by two chains inside the wheels to the axle of the leading wheels, with another

**When Radstock's** Sentinel engines were unavailable two ex-Lancashire and Yorkshire Class 21 0-4-0ST engines were used as substitutes – Nos.11202 (later 51202) and 11212 (later 51212), the former being attached to the S&D at least three times over the years. They were two of a class designed by John Aspinall and team, with fifty-seven built at the Horwich Works between 1891 and 1910. Engine No. 11202 was one of the first five built in 1891 and the picture to the left is one of her four sisters. They seem to have been a very effective class of locomotive and their brief sojourns to Somerset seem to have confirmed this view. (Author)

coupling chain connected to the trailing wheels. Steam is supplied by a vertical boiler of a traditional Sentinel pattern in the cab, whilst a 600 gallon water-tank occupies the middle space between the engine and the boiler. The exhaust from the engines is passed through the superheater's coil, which is situated above the boiler tubes and then discharged through four nozzles on the cab roof. Their squat appearance and the general "packing in" of parts are due to the sharp curves they have to traverse and the limited height clearance of 10ft 5in at Tyning Bridge.'

These engines proved very effective in service, but the enclosed cabs were reputed to be most uncomfortable during hot weather; this problem was probably outweighed by their strength, reliability and the annual savings they achieved over their predecessors. However, with only two available, there was no natural reserve during absences and substitutes had to be found from elsewhere in the LMS's fleet. When both were undergoing maintenance, ex-Lancashire and Yorkshire 1891 built 'Pug' Class 0-4-0Ts, Nos. 11202 and 11212, took over. These game old engines would return to Radstock at least three more times before being withdrawn from service seven years later.

Although the third group of engines to be acquired as a result of the 1928 review were three more superheated 4-4-0 tender engines, it was a decision only reached after trials of other types of locomotives were held between 1925 and 1927. Two of these involved a Deeley Class 990 4-4-0 and a newly introduced Horwich Mogul, in this case engine No. 13064, built at Crewe in 1927.

The Deeley engine, No.995, offered nothing new to the S&D's footplate crew and was noted for burning more coal than the other 4-4-0s in service. This temporary allocation was probably no more than a way of finding a home for members of this class. By 1925, they had largely been superseded by newer and better engines on the LMS and were struggling to find a home. With no success on the S&D, all ten of the class were condemned and by 1928 had been reduced to scrap. The Mogul was a different kettle of fish, though. Engine No. 13064 had only just been built at Crewe in 1927 when assigned to Highbridge in November that year.

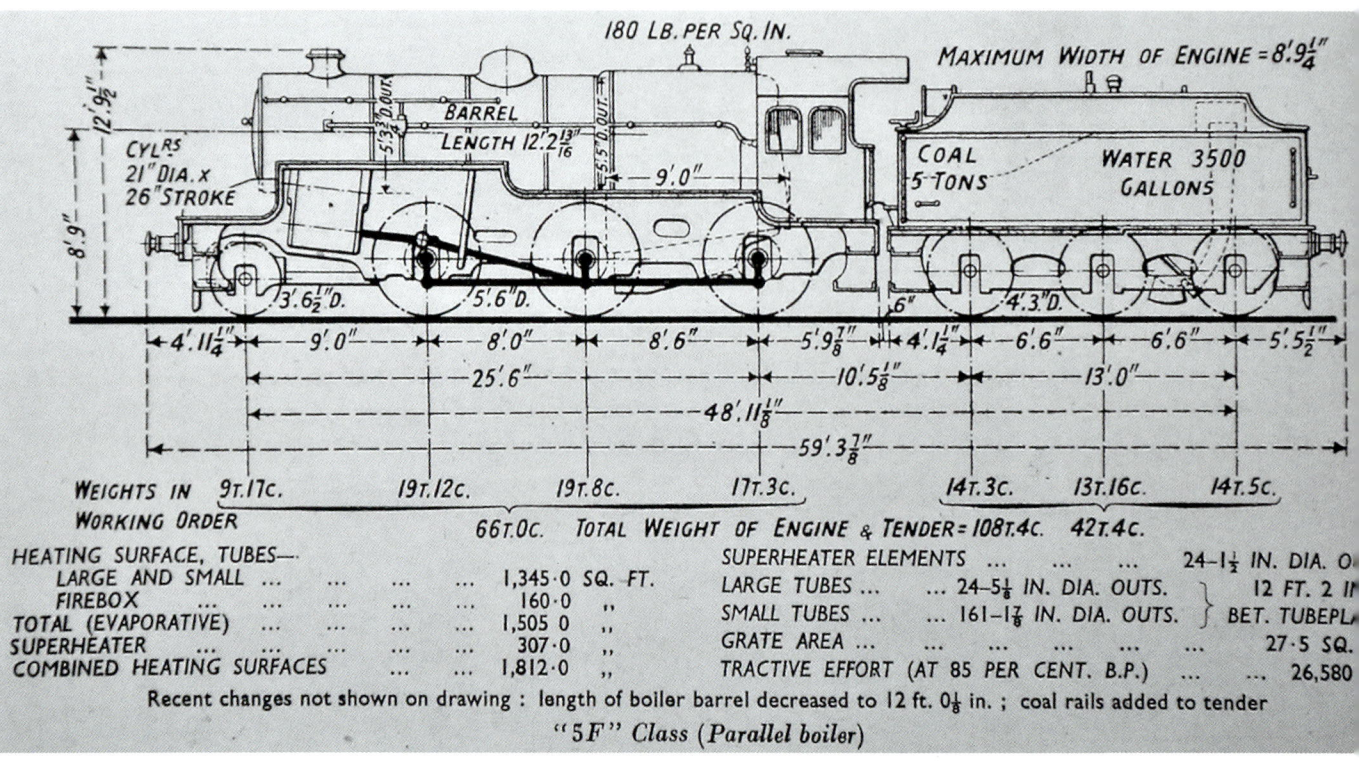

**The impressive** lines of the Hughes designed Horwich built 2-6-0s are captured well in the diagram and the photo presented here, the latter taken at the works in 1926. This engine was the first of 245 constructed by 1932. This powerful class of engine would have offered the S&D a much-needed boost to its locomotive fleet, but only one, No. 13064, was assigned to the line, albeit for trials only. However, even this short loan gave the footplate crew a tantalising glimpse of what might soon be available at Derby, Crewe and Horwich. The coming decade would reveal three of them – Stanier and Coleman's Black Five 4-6-0s, the 8F 2-8-0 and the 3P 2-6-2 tank engines. (Author)

The Mogul's six coupled wheels and their reserves of strength offered Archbutt something a little extra, particularly when rostered to passenger duties on which it performed well. When assigned to freight work, it was slightly less commanding, but effective nonetheless. Following these successful trials there would have been a natural expectation that a permanent allocation might soon follow. But it was not to be, probably because the LMS needed them all for their own services. After this one can only assume that there was some disappointed footplate crew when three more 4-4-0s arrived during 1928 instead. The fact that they were the latest product of the Fowler/Chambers partnership probably held little water.

They were in fact modified 483 Class locomotives, the first of which were built at Derby in 1928, with a total reaching 138 during 1932. Initially, they were priced at £5,000 each, but this was reduced to £4,320 when a compensating saving was taken into account, which placed a scrap value on ex-S&D parts removed from other engines when undergoing repair.

The new engines, given the designation 2P, were originally to be fitted with outside cylinders and inside Stephenson valve gear. This was an attempt to correct what was thought to be a weakness in the cylinder design of the earlier engines which, according to a

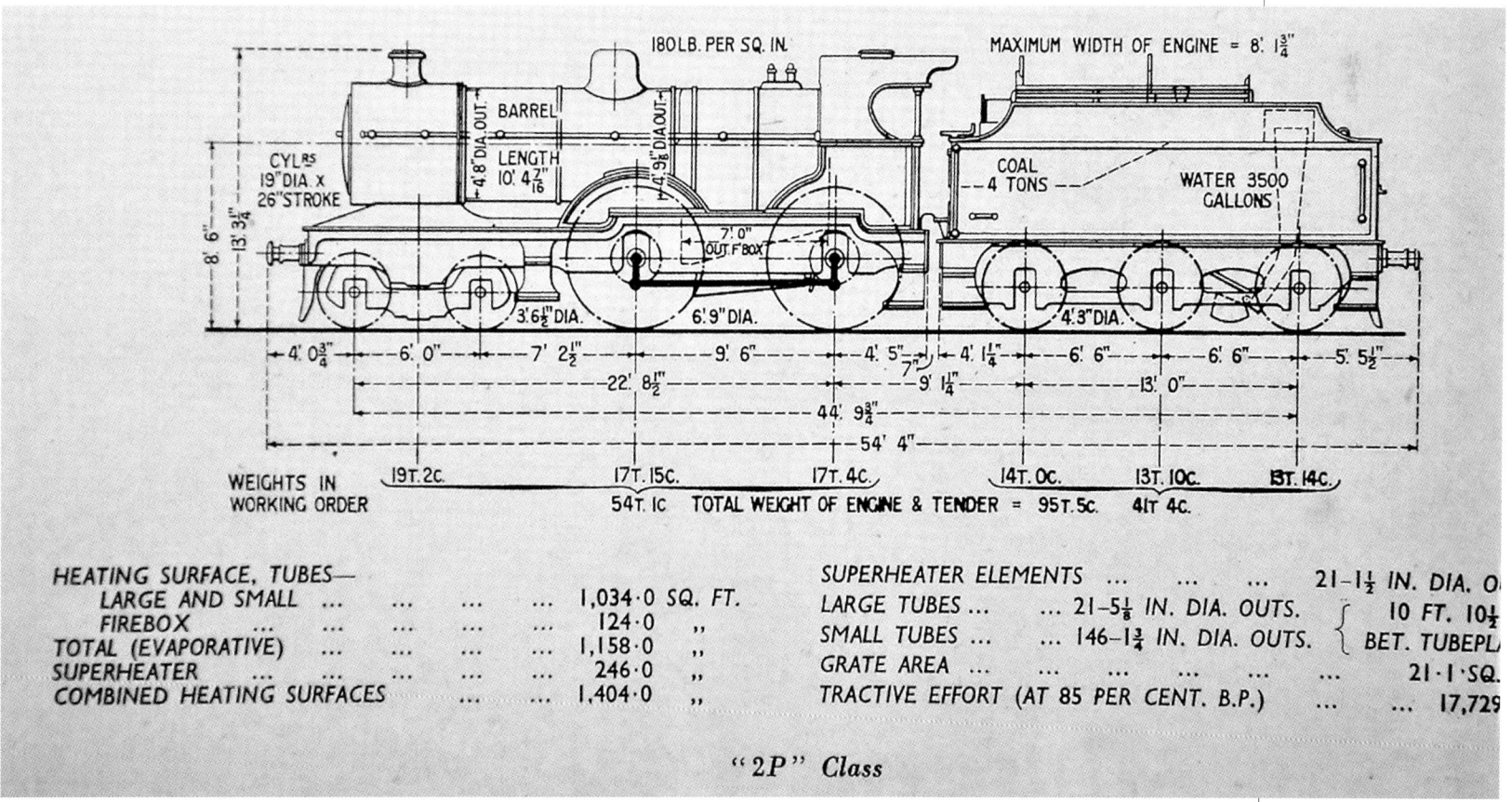

"2P" Class

**Fowler Class** 2P 4-4-0s were something of an anachronism when they appeared between 1928 and 1932 in that they reflected a policy and way of thinking more suited to a bygone age. By comparison with locomotive design in the hands of draughtsmen with the GWR, LNER and SR, the LMS seemed to advance more slowly at this time. One hundred and thirty-eight more 4-4-0s seemed to exacerbate the problem. However, things were slowly changing and the purchase of Garratt 2-6-0 + 0-6-2 mineral engines in 1927 and the appearance of fifty 4-6-0 Royal Scots, mostly designed and built by North British Locomotive Company in the same year, showed a growing willingness to experiment. In the meantime, the S&D soldiered on with its remaining 15 or so 4-4-0s. (Author)

**The first** of the S&D's three new Class 2Ps awaits delivery to Somerset in the LMS's Roundhouse. at Derby during June 1928. It would be accompanied south by sister engine No. 45 and No. 46 a month later. Even in a black and white photograph, the quality of its blue paintwork is only too apparent. In due course they will be renumbered 633, 634 and 635 and remain in service with BR until 1959, 1962 and 1961 respectively, but not the S&D, which they left in the 1930s. (R. Hillier)

contemporary report, 'affected their performance somewhat'. However, this solution seems to have created loading gauge problems itself with clearance being a problem on some secondary routes. This led to a rethink in the drawing office. As a result, the valves remained underneath the cylinders, which in turn had their diameter reduced in size by one inch to 19in, with the stroke remaining at 26in. This meant that the Stephenson gear didn't have to be altered and retained short travel valves.

**In November 1933,** the first of the S&D's three 2P engines was fitted with a Dabeg feedwater heater and pump, as witnessed in this photograph by the rather unsightly equipment added to 653's front end. This system was developed in Austria and came to the attention of Nigel Gresley who began trials with it in 1925. Two LNER O2 2-8-0 engines were involved, Nos. 3499 and 3500, but it was found that the Dabeg system 'presented no advantage compared to steam injectors' so went no further. However, the LMS, believing that some benefit might be gained, equipped an ex-L&Y Class 8 4-6-0 four-cylinder superheated engine during 1927 in an effort to improve the mediocre performance of this 1908/09 built class. A 2P, No. 706, was similarly equipped in 1931 and ran a series of test runs between Derby and Bristol. These identified savings in the order of 11.5 per cent for coal and 14.8 per cent for water after a new spray pipe had been fitted (2.48 per cent and 471 per cent before). Overall, the engine's performance was satisfactory, and it was confidently predicted that 'preheating and de-aerating of the feedwater in the heater might prove an advantage as regards reducing scale and corrosion in the boiler'. Tests continued and engines 633 and 653 were absorbed into the programme, but it appears that any benefits derived from the pump were too small to warrant the expenditure of taking the experiment any further and work came to an end. (D. Neal)

**This picture** taken by Tynings Bridge at Radstock in 1929 sums up the problems faced by the S&D especially in the summer season when heavy passenger trains passed over the network. 0-6-0 No. 60 and new 2P 4-4-0 No. 45 have to double head a heavy express service. A Horwich Mogul or a Stanier Black Five may have operated alone and saved the cost of a second engine. (D. Neal)

G7S boilers were fitted, but the working pressure was raised from 160 to 180lb. and following the pattern set by the later 2-8-0s Ramsbottom safety valves gave way to the Ross Pop version. Other changes included a reduction in the size of the coupled wheels used on the 483s from 7ft 0½in to 6ft 9in and lower chimneys and domes. In addition, there was a slight reduction in the combined heating surfaces from 1,170sq ft to 1,158sq ft, with a superheater of 243sq ft compared to the 313sq ft of the early 483s which helped produce 17,729lb. of tractive effort down from 22,649lb. The tenders were straight sided and carried 4 tons of coal and 3,500 gallons of water, 250 gallons more, and their overall weight at 95 tons and 5cwt was 4 tons 18cwt greater than the 483s.

As these locomotives were part of a seventy engine order for delivery to the LMS in 1928, when notified of the S&D's requirement three were simply taken from this programme. Their numbers, 575, 576 and 580, were changed to 44, 45 and 46 in which state they arrived at Highbridge in June and July that year. Shortly afterwards, engines Nos. 67 to 71 were renumbered 39 to 43 to provide some sequential order.

However, these were changed again in 1930 when 39-43 became 322 to 326 and 44-46 became 633 to 635. As things turned out, all these superheated 4-4-0s didn't remain long on the S&D and until Black Fives arrived, the LMS rotated the superheated 4-4-0s. So, 322, 323 and 324 were transferred to Millhouse and 325 and 326 to Saltley in June 1936. Then, in mid-1937, the final three, Nos. 633, 634 and 635, moved to Gloucester, with replacements arriving in the form of 2Ps 600 to 602, 699, 700 and possibly some others. In due course, 326 and 634 and No. 700 (on a temporary basis) did return to the S&D and became a common sight until they were withdrawn from service in 1956 and 1962.

In assessing their overall performance, it is hard not to question their construction at a time when such small engines were falling out of favour on main line passenger duties. Although 4-4-0s were good in their day, their time was passing and in a changing world they were increasingly consigned to light secondary route duties or double-heading as loads increased. In retrospect, Brian Haresnape, in his book *Fowler Locomotives*, probably summed these engines up best when he wrote:

> 'The 2Ps were an indifferent class due to poor front end layout…very little change was made to the appearance of these engines throughout their somewhat undistinguished careers. They never shone in terms of performance on the road. On the credit side, they were simple and robust engines whose average repair costs in pence per mile were remarkably low.'

The arrival of these engines, along with the five new 2-8-0s and the seven 0-6-0 Jinty 3Fs, brought to a conclusion the replacement programme agreed in May 1928 which assumed the scrapping of twelve of the older locomotives. The disposals were soon well in hand and took with it the last of George England's 1860s engines, No. 28A which went in April 1928 after sixty-seven years of service. In fact, it was withdrawn possibly in anticipation of the May agreement, so may well have been part of the compensatory allowance agreed by the LMS. Likewise, Vulcan built 0-4-4T No. 52 and Fox, Walker 0-6-0T No. 8 both went in May. These were soon followed by five of the John Fowler 0-6-0s, Nos. 20 to 24, number 19 having been scrapped in June 1927, two Derby built 4-4-0s, Nos.15 and 16, four Vulcan 0-6-0 tender engines, Nos. 25, 26, 27 and 56. And to this list were added all three Radstock 'Dazzlers' in 1929/30. Nevertheless, this wasn't an end to the changes, because post-1930, the programme became the sole responsibility of the LMS.

It was perhaps unavoidable that the financial problems that had beset the S&D for so long should re-surface again and drive the need for major changes in an effort to stave off bankruptcy. No matter what was tried, it seems to have made little difference and the trend continued into 1929 when expenditure exceeded income by nearly £12,000. One bad year might have been forgotten but the cycle of loss was now a continuous one in which fresh economies failed to halt the decline and in 1929 the picture was made worse by the Wall Street Crash and the deep recession that followed. Some reforms were swiftly introduced in an effort to stem the haemorrhage and these saw any last vestige of independent management come to an end on 1 July 1930. As a result, the Traffic Superintendent's office at Bath was shut and its staff either transferred to the LMS or made redundant. This was preceded by the closure of the works at Highbridge in May with the loss of 300 or so jobs. From this point on, day to day operations passed to Derby where the railway was managed on behalf of the Joint Committee. At the same time, the Southern Railway took over responsibility for the infrastructure. And so the posts of Locomotive Superintendent and Resident Engineer became redundant. Shortly after these changes were announced, the Derby born Archbutt finally returned to the city to see out the rest of his career with the LMS. He retired in the early 1950s and passed away at his home in Harrow in 1967.

Inevitably, any company in straitened circumstances will make cuts in the hope that this might reduce the loss and turn things round. In 1933, the S&D's shipping interests were wound up. One cargo vessel was sold in 1934 and the other scrapped a year later. However, the wharf at Highbridge remained in service until 1950. Elsewhere, the Wimborne line was closed except for a short section, and there were a series of other minor cuts in men, equipment and facilities. Whilst all this was happening, a change of some significance was taking place away from the S&D, which would in time have a strong influence on its locomotive fleet.

On 1 January 1932, William Stanier became the LMS's CME, bringing with him many years of experience working for the GWR, under George Churchward and then Charles Collett. Although not a designer of any note himself, he was a man full of ideas and someone not easily swayed from a course of action he deemed essential. But with James Anderson still holding sway over locomotive policy as Motive Power Superintendent, there seemed every chance of there being a battle ahead. However, this was averted by Ernest Lemon, recently promoted to Vice-President in charge of Railway Traffic, Operating and Commercial. He spotted the danger Anderson posed and 'relieved him of his duties with immediate effect on 31st October 1932', replacing him with David Urie, who fostered a more informed view of the future than his predecessor.

Perhaps of even greater importance to the S&D, and railway history in general, was the

**This picture** captures engine No. 28A during 1928 in a derelict and partially dismantled state at Highbridge awaiting a trip to the smelters. It was a sad end to a locomotive that was one of the 1861 original intake to grace the old Somerset Central Railway. (Author)

**Stanier and** Coleman's maid of all work and a classic locomotive design, the 1934 introduced mixed traffic 4-6-0 5F, known far and wide as the Black Fives. Before becoming the LMS's Chief Draughtsman in 1935 Tom Coleman was Chief Draughtsman at Crewe where these engines were designed and some of the 842 built before production ended in 1947. He continued to oversee modifications to the design until he retired in 1949. (Author)

**Key members** of the LMS's CME Department in 1936 when meeting at a retirement celebration for George Shawcross, the Mechanical Engineer, at Horwich. Far left – Robert Riddles, who would go on to become British Railways' first CME, though the title was slightly different, and be responsible for a standardisation locomotive programme that would affect the S&D as it did the rest of the network. Third left – Henry George Ivatt, who became CME in 1945. Fifth left – the dominating figure of William Stanier and, half hidden behind him, Tom Coleman. Third from the right – Frederick Lemon, the Works Superintendent at Crewe and was responsible for producing many of Stanier's most iconic locomotives, including the 5F 2-6-0s, 4-6-0 Jubilees, Black Fives, 8F 2-8-0s and the Coronations. On the far right stands Herbert Chambers, who by this time had become the CME's Personal Assistant, a post he occupied until his death in September 1937. (Author)

appointment of Tom Coleman as the LMS's Chief Draughtsman for Locomotives and Rolling Stock in 1935, succeeding the ailing Herbert Chambers, who moved to Euston to become Stanier's Personal Assistant. Although the outgoing CD had been remarkably successful in serving Fowler, Lemon and then Stanier in producing many good locomotives, he seems to have been exhausted by the effort and had probably lost much of his energy and ambition in the process. It didn't help that illness was taking an increasingly heavy toll on him, making it even harder to lead such a large team at such a most dynamic moment in time. And one thing is certain, if Fowler had been active in the design field Stanier was even more vigorous. It was as though the end of his years as an assistant to the GWR's CME acted as a starting gun for the release of his pent-up ambitions and ideas. However, the changes he wrought didn't happen overnight and new engines were almost trickle fed into the system from 1933 onwards. First came his twelve Princess Class Pacifics, then forty mixed traffic 5F 2-6-0s and, during 1934, the first of 190 Jubilee Class 4-6-0s that were built by 1936. Then came a class of the greatest significance, the 4-6-0 Black Fives and it was here that the Coleman influence first came to bear in a significant way.

Coleman was born in Gloucestershire in 1885 and was brought up there and in Surrey before his family moved to Endon north of Stoke. Having demonstrated a gift for mathematics, model making and engineering, he joined the locomotive builders Kerr, Stuart on a five-year apprenticeship in 1900. On completion, he found employment with the North Staffordshire Railway, rising to become Chief Draughtsman in the immediate post-war period. Along the way, he assisted or led in designing many locomotives, demonstrating the depth of his talent and grasp of design issues. Grouping in 1923 did not have an immediate impact on his post in Stoke, but within two years he found himself despatched to Horwich as Chief Draughtsman and began the process of cementing his position in the company. Here he remained until 1933, when he was posted to Crewe as CD by Stanier, and it was here that his influence began to dominate design work across the LMS. It was at this point that the Black Fives went from specification to being an active design then reality, with Coleman and his team at Crewe taking the lead in this work.

Although the initial batch of these versatile engines were constructed in 1934, it would be three years before they became a permanent feature on the S&D. During this period, the main line passenger services, at their heaviest in the summer months, remained in the charge of the 2P 4-4-0s, with the 7F 2-8-0s pressed into service when demand was high. As *The Locomotive Magazine* reported at the time, this holiday traffic to and from the south coast 'attains dimensions which fully tax the capacity of the line, as many as 24 passenger trains being booked in the southbound direction on a busy summer Saturday, including eight out of Bath between 1 pm and 3 pm'. With the volume of traffic increasing, and with it the size of the trains, the LMS's Bristol Motive Power Depot,

**One diagram** doesn't do justice to the variations made to the Black Five design over its seventeen-year construction programme. However, this picture does at least record many of the principle dimensions and gives a basic outline of this class of 842 locomotives. (Author)

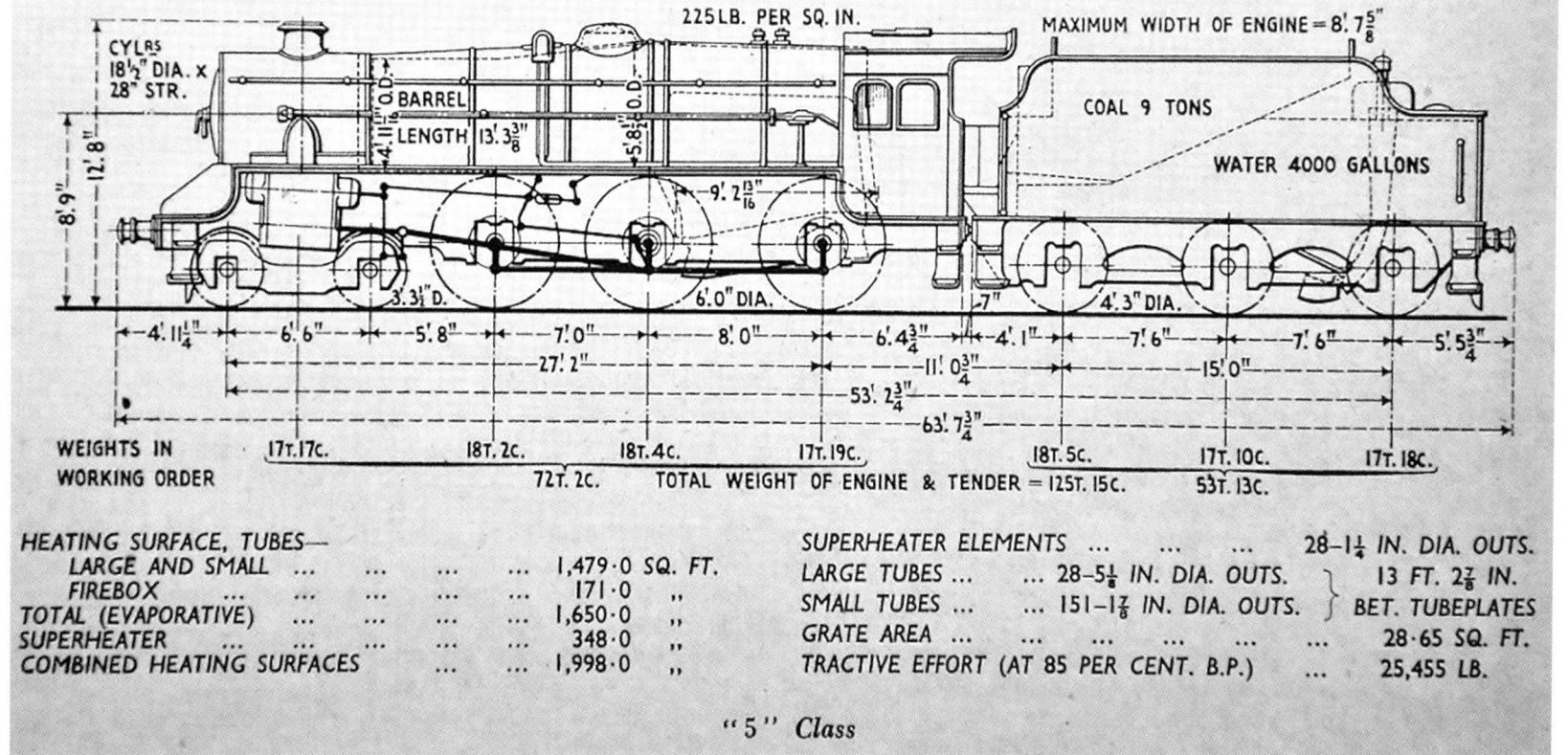

**In 1938,** Black Five No. 5432, preceded by No. 5228, one of 150 built at the Vulcan Foundry in 1934/35 and 276 constructed by Armstrong Whitworth between 1936 and 1938 respectively, arrived on the S&D to begin trials. In May, with these tests completed, No. 5432 remained with the S&D and ran the 10.20 am Bath to Bournemouth semi-fast. On the same day sister engine, No.5440, another recently produced Armstrong Whitworth engine, pulled the Pines Express unassisted to Bournemouth. In so doing, it cemented their relationship with the S&D and the class remained a regular feature on the line for many more years. It appears that engine No. 5432 was one of the last of two Black Fives to work on the S&D, eventually being withdrawn in 1958.
(D. Neal)

which now controlled operations over the line, found it increasingly difficult to provide locomotives of adequate power and so reduce the requirement for double-heading.

It was against this background that Black Fives were trialled over the line, but not until March 1938, when work to strengthen the river bridges between Mangotsfield, on the eastern edge of Bristol, and Bath, to take these 125 ton plus locomotives, had been completed. This allowed engine No. 5228, shortly followed by 5432, to advance safely to Bath to begin work. Part of these tests was to determine how much these engines could haul without assistance over the steep inclines they faced. This built on knowledge acquired over the years when operating other classes of locomotives on passenger trains. This resulted in a simple performance chart covering three of the difficult sections of track ready for the crew of the six Class 5s that worked the line that summer. This aid was later published in the railway press and is produced in part on the adjoining page:

As this chart reveals, there were obvious advantages in acquiring the Class 5s, but it didn't remove the need for double heading completely, especially in the summer months when loads grew ever larger. This was particularly so on the Bath-Masbury section, which was a severe test for any locomotive. So, it was a partial solution, not a complete answer; nevertheless, these engines were a great improvement and soon became a permanent part of life on the railway.

While high speeds over the line were difficult to achieve, and rarely exceeded 60mph, a good average was essential if schedules were to be met. The Black Fives were more than capable of achieving this and regularly performed runs of a 'meritorious character', as *Locomotive Magazine* described

| | Bath-Masbury (a) | Masbury-Corfe Mullen | Corfe-Mullen-Bournemouth W |
|---|---|---|---|
| Class 2 4-4-0 ("600 type") | 210 tons | 315 tons | 280 tons |
| Class 3F 0-6-0 | 190 tons | 290 tons | 260 tons |
| Class 4F 0-6-0 | 240 tons | 365 tons | 320 tons |
| Class 5 4-6-0 | 270 tons | 405 tons | 380 tons |
| Class 7F 2-8-0 | 310 tons | 450 tons | 415 tons |

**Note (a)** – In the northbound direction, the figures in the tonnage column apply to the uphill section Evercreech Junction-Binegar; between Binegar and Bath northbound trains are permitted the tonnage shown above Masbury-Corfe Mullen, with a single engine. The loadings shown in the second column apply in the northbound direction from Corfe Mullen to Evercreech Junction, whilst those for Corfe Mullen-Bournemouth West apply in both directions. The provision of tonnage ratings for the ex-S&D engines now classified as F will be noted, but their use on passenger trains has mainly been confined to emergency and occasional heavy but easily timed excursion trains.

**Black Five** No. 5432 again photographed in her early days on the S&D, on this occasion pulling the eight carriage up Pines Express, apparently with great ease as the engine doesn't appear to be labouring in any way. The fireman's arm can be seen resting on the window's edge suggesting his engine is working well and he can take a 'breather' before the next steep climb begins. (D. Neal)

**An exceptionally** powerful mix of engines for a heavy summer express on the S&D. Although very strong, the Black Fives with loads creeping over 400 tons might need assistance over very steep sections of track. In this case, the 4-4-0 seems to be working much harder than the 5, if the smoke is anything to go by. The number of the first engine isn't very clear though may end with '32' which suggests it was either passing through or was only temporarily with the S&D. The Black Five's number is illegible. (D. Neal)

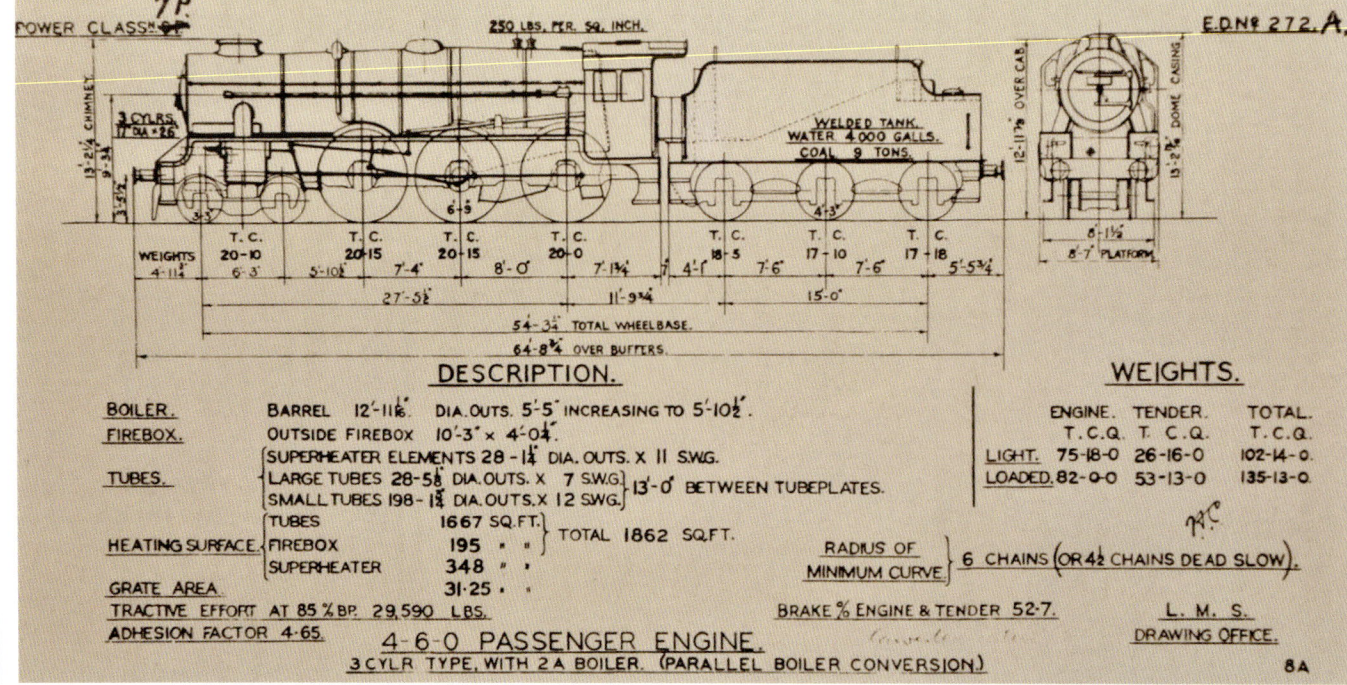

**The Jubilee** Class 5XP as built between 1934 and 1936. These engines enjoyed a long career, and some would continue in service into 1967. Their performance, considered disappointing when built, underwent modification in the hands of Tom Coleman, whose initials can be seen on this diagram. (Author)

their work. Then the author added details of one particular run at this time with engine No. 5432 when heading south with a Pines Express weighing 344 tons:

'It covered the 16.1 miles from Templecombe No. 2 to Blandford in 23 minutes, inclusive of three slacks for crossing loops, and a permanent way slack, and the 10.9 miles from Blandford to Broadstone in 16¼ minutes.'

By comparison, the 2Ps could only achieve comparable times when hauling loads significantly less at 200 or so tons and more often than not with a second engine providing assistance.

The Black Fives weren't the only LMS 4-6-0s to reach the S&D. On a number of occasions, three-cylinder Class 5XP Jubilees worked express trains into Bath from the north but could go no further due to weight restrictions. These engines were developed by Stanier from Fowler's Patriot Class which first appeared in 1930. Construction ran on until 1936, by which time the new CME was toying with ideas for a modified version and took the last five on order and revised them in accordance with his own design philosophy. As a result, they were built with redesigned axleboxes and wheels, low-degree superheat domeless and tapered boilers, modified cabs and tenders. However, they proved to be something of a disappointment, which in due course was shown to be caused by, amongst other things, the 14 element superheater and problems with draughting. When modified they were deemed to be *'competent locomotives which performed their role of intermediate express power efficiently'*. The one hundred and ninety built by 1936 worked over all four of the LMS's divisions, but due to their size never achieved the route availability of the more versatile Black Fives.

As the Black Fives began to make their mark, the first of Stanier's two-cylinder Class 3P 2-6-2 tank engines reached the S&D. Initially three engines were allocated in 1938, Nos. 179 to 181,

**Jubilee No.** 5660 *Rooke* became a regular visitor to Bristol Temple Meads Station and is captured here in the late 1930s with a train from the north. In time, Jubilees would be turned and then pull such trains to Bath where another engine or engines would pull the express over the Mendips and on to Bournemouth. (D. Neal)

**Tom Coleman's** initials again appear on this diagram, suggesting his proprietary interest in the class as the LMS's Chief Draughtsman, which he became in 1935, the year these engines first appeared. On paper, the design looked sound and sought to improve upon Fowler's 2-6-2Ts, but in practice they were seen as under-performers which modifications failed to correct. Nevertheless, they served the LMS and BR until the early 1960s. *(Author)*

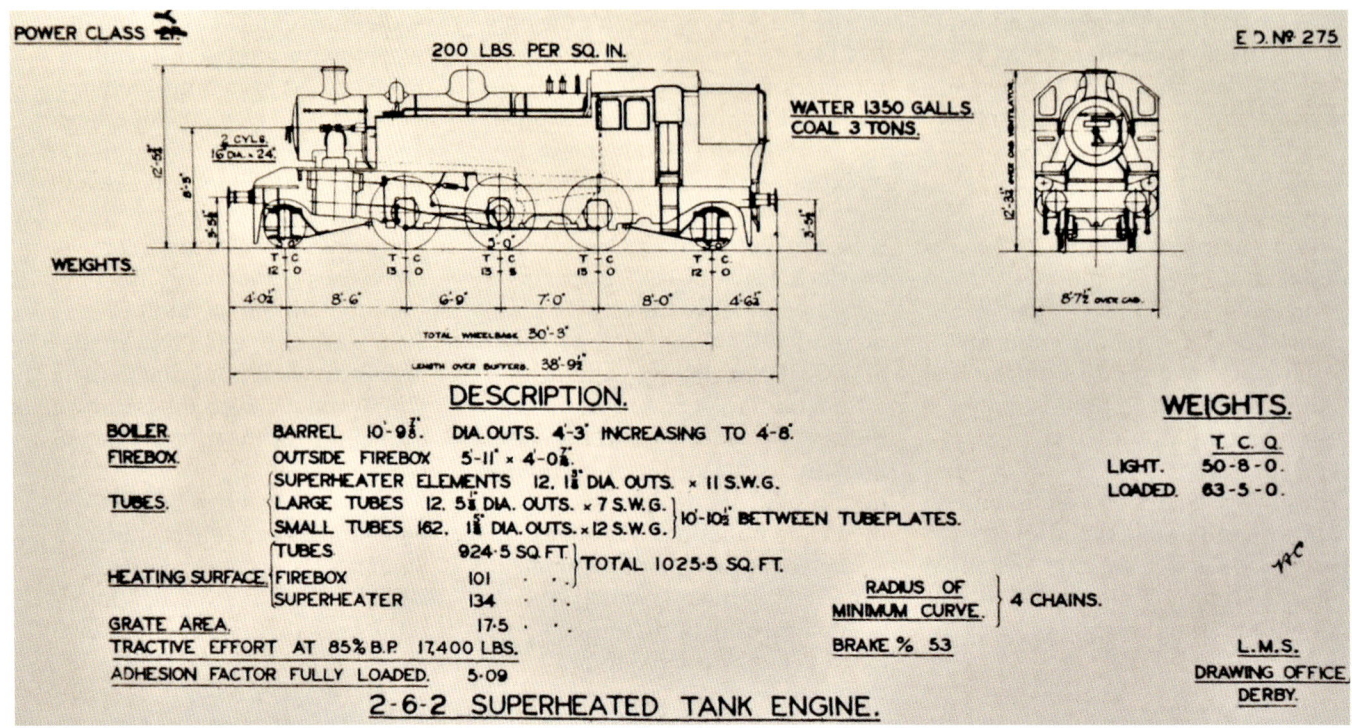

and these found ready use pulling passenger trains daily from Bath to Templecombe and back. To this task was added banking duties between Bath Junction and Combe Down tunnel and hauling local trains to Bristol. These locomotives were once again derived from a Fowler design, on this occasion one that appeared at the end of his career in 1930 with construction of seventy running on into Stanier's tenure. Once again, the key change focussed on the installation of a tapered boiler and modifications to the cab. In this state, 139 were built between 1935 and 1938 and found wide use across the network. However, they, like the Jubilees, proved to be undistinguished performers, but unlike the 4-6-0s various modifications failed to improve their reputation. This may be why some found their way to the Somerset and Dorset, other users looking for something better in their place.

Stanier's third engine to have an impact on the S&D also appeared in 1935, when the first batch of twelve 8F 2-8-0s rolled out of the workshops at Crewe, but they didn't see service on the S&D until 1941. However, it is rumoured but not confirmed that an example may have visited the line in 1938 when the bridges between Bristol and Bath had been strengthened. These heavy freight engines could be said to be the result of many years of fruitless efforts by Hughes and Fowler who had both been eager to see a large locomotive such as this making its way into the LMS's fleet. We have observed Hughes' unsuccessful attempts to build a 2-8-2 and Fowler's failure to build on the success of the S&D's pre-war 2-8-0s. But he did try and had worked hard to get his ideas into the ARLE's sphere of activities, then attempted the same with the Midland Railway. Finally, he moved the LMS closer towards his developing model for a goods engine, only to see his efforts resulting in the 7F 0-8-0, the large and expensive Garretts, the single 0-10-0 Lickey Banker and five more 2-8-0s for the S&D. It was left to Stanier and Coleman to realise these long-held ambitions and produce a locomotive of modern design, big enough to meet the company's growing freight needs.

When the first batch of twelve appeared, these 2-8-0s were given the designation 7F, but this was soon upgraded to 8F, reflecting the fact that they were a step up from the Fowler 2-8-0s and capable of much more. With an overall length

**Derby constructed** 2-6-2T No. 76 as it appeared in 1935 when built. Despite their good pedigree, they were considered to be, as reported at the time, 'indifferent performers'. Rebuilding was considered in the post-war years and a number received larger boilers, but these appear to have shown no appreciable improvement over the other engines for the cost involved, so the project came to nought. (R. Hillier)

**In 1932** the LMS extended a new numbering system to locomotives on the S&D. In this case, 9675 became 13805 and, according to notes with this picture, is seen at Derby following a period of general repair just prior to the war. The engine has just received its black top coat and the tender still appears to be in primered condition. The buffer beam is also grey awaiting a final coat of paint. (Author)

**The 8Fs** proved to be remarkable engines and by the time production ended in 1945, 852 had been built. The timing of their arrival could not have been better planned with war so close. During the conflict, they found wide use on networks at home and overseas and were built under licence by the GWR, LNER and Southern Railway, such was their importance to the war effort. (Author)

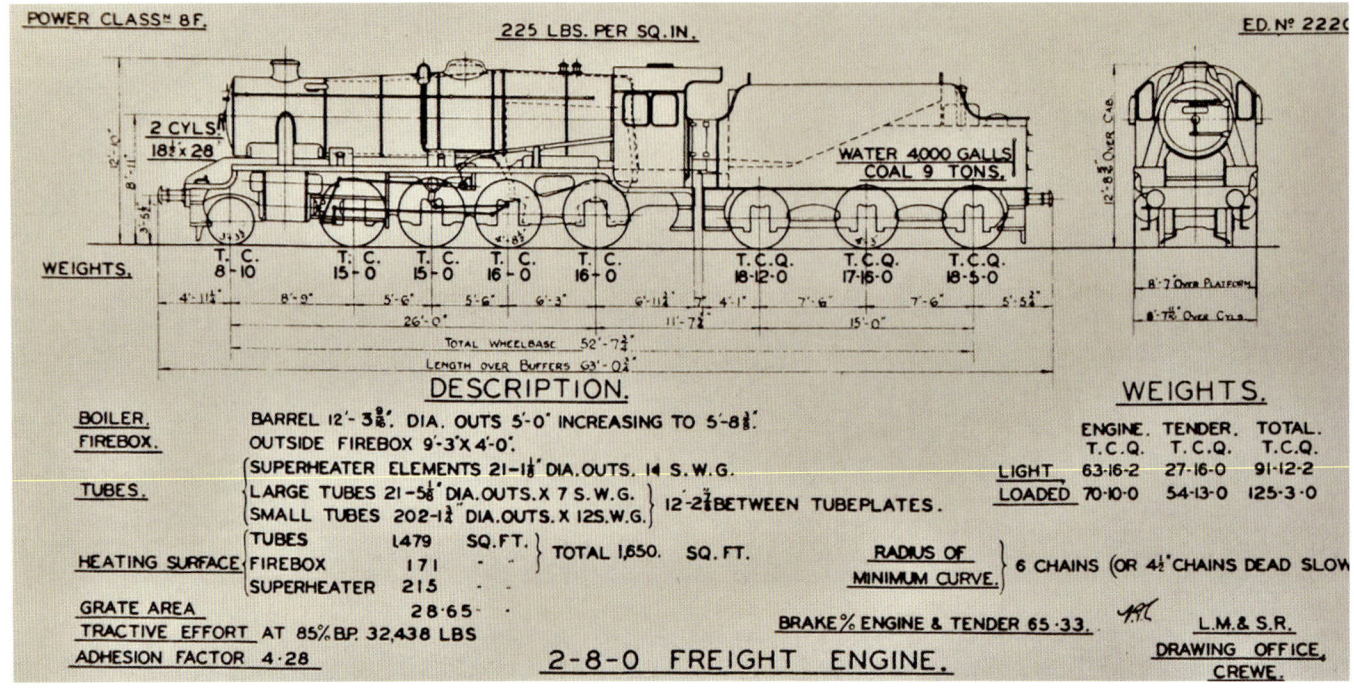

of 63ft 0¾in and a total weight of 125 tons 15cwt they were slightly more than 7ft 2in longer and 37 tons heavier than the second batch of 2-8-0s built for the S&D in 1925. However, the initial batch of Stanier engines were fitted with domeless taper-boilers with a grate area of 27.8sq ft and a total evaporative heating surface of 1,463sq ft. – a tube surface of 1,308sq ft, plus a firebox surface of 155sq ft. To this was added a superheater with a surface area of 230sq ft. Under Coleman's guiding hands, these boilers were soon replaced by a version with separate dome and top feed, plus a firebox containing a sloping rather than a vertical throatplate, which produced a total evaporative heating surface of 1,650sq ft. This revised throatplate arrangement had a number of benefits. It increased firebox volume and lengthened the path of the flames before entering the tubes so improved combustion and, at the same time, reduced the weight of the boiler.

Other than these changes, the basic layout of batch one and later engines remained the same. There were two outside cylinders measuring 18½in by 28in, as against the Fowler engines' 21in by 28in, and Walschaert long travel valve gear. The coupled wheels had a diameter of 4ft 8½in and the Bissel truck 3ft 3½in wheels. Meanwhile, the tender dwarfed the version attached to the S&D 2-8-0s and were built to contain 9 tons of coal and 4,000 gallons of water. Their cabs also reflected Stanier's views on design. They were well laid out, enclosed and reflected quite advanced views on ergonomics in the way they matched equipment to the capabilities of its users. Working on the footplate of a steam locomotive was never going to be a comfortable experience, but it could at least be made more tolerable.

The slow production rate of these new, more powerful freight locomotives meant that staff at all motive power depots, where demand for such engines was high, faced a frustrating wait for them. This would have increased exponentially in the summer of 1939 as the country mobilised its military force and engaged its industrial might. It is little wonder then that it would take until 1941 for three to turn up in Somerset for a short while to be based at Bath, to, as *Locomotive Magazine* reported, 'supplement the S&D Fowler 2-8-0s on heavy goods trains'. But it proved to be a short-term loan, or so it seems, with these engines soon being re-absorbed by the LMS further north for war service. In fact, early in the conflict the War Department, ever aware of the

**Production of** the 8Fs at Crewe began in a leisurely way – twelve in 1935, three the next year, twelve in 1937, two in 1938 and then a sudden increase to twenty-eight on the eve of war. However, Vulcan Foundry supplemented this slow rate by building sixty-nine in 1936/37, so giving the LMS a substantial number to play with as the conflict bit deep into the country's resources. From here, the construction rate increased rapidly. In this picture, engine No. 8013, one of the second batch of engines, nears completion at Crewe in December 1936. (R. Hillier)

**Engine No.** 8103 was part of a batch of 128 8Fs which were built at Crewe in 1939. This picture captures the look of these engines when new. However, heavy wartime usage and shortage of maintenance staff and cleaners, as many entered the services, soon reduced these work horses to a shabby state. Nevertheless, the robustness of their design made them rise above their deteriorating condition to provide effective service during the war. (D. Neal)

**The works** at Eastleigh were responsible for building twenty-three 2-8-0s during 1943/44, including engine No. 48660 pictured here in 1948 at Birkenhead with its nationalised number, but before BR markings were applied. Towards the end of steam, this engine found its way to the S&D to replace the 1914 built 7Fs which were reaching the end of their lives. (R. Hillier)

**A supreme** effort by the railway was essential, by companies both large and small, if war needs were to be met effectively. In 1942/43 this was celebrated by this simple and evocative painting which was circulated widely at the time. This, as the picture makes clear, was the requirement for only one of a number of campaigns in Europe and others far beyond its boundaries. (Author)

**An Army moves**

440 Special Troop Trains and 1,150 Freight Trains were needed when our Army left for North Africa

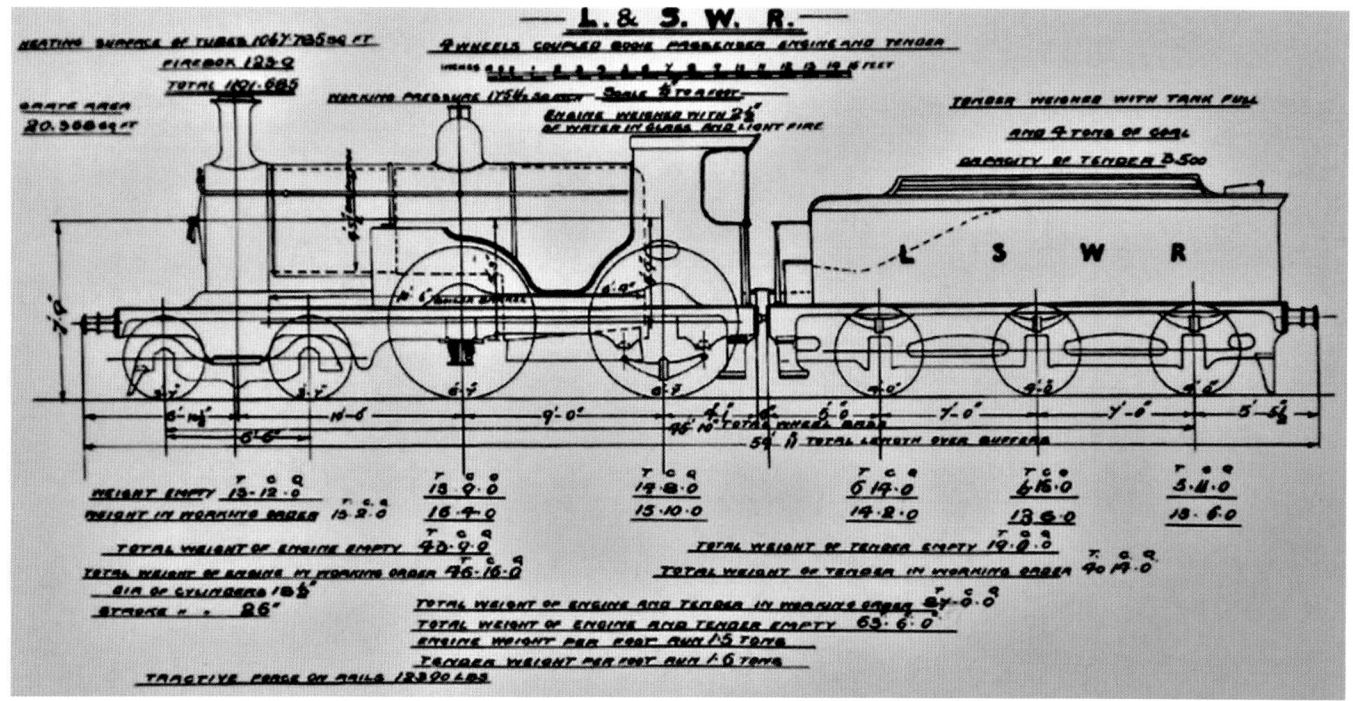

important role railways played in supporting the front line in the First World War, soon considered future need. There were a number of options to assess and in 1940/41, the 8F was chosen as one way of meeting this need. However, before production could be ramped up and new engines supplied, the WD requisitioned fifty-one locomotives from existing stock and shipped them abroad. It was little wonder that the S&D only enjoyed their presence for such a short time. But they would return, some simply passing through during the war and after, and then, in 1961, on allocation to Bath on a permanent basis. Either way, their existence provided a hopeful pointer to a future beyond the war. In 1939/40, with battlefield disasters followed by isolation, mass bombing and the threat of invasion, this was a future few could view without

some pessimism. Survival was all that mattered and to do this, the railways were forced to work beyond capacity, with staff, locomotives, rolling stock and infrastructure continually hard pressed, to meet all demands placed on them.

Day to day descriptions of life on this line during the war years are rare, particularly when recording the movement of heavy

**The exigencies** of war meant a number of 'foreign' engines would have to be accommodated to ensure that all traffic needs could be met. The LSWR/SR T9 4-4-0s would have been a familiar sight to S&D crew before the war as they plied their trade west of Salisbury, occasionally making their way along the line to Bath. But these appearances increased during the conflict. Designed by Dugald Drummond and his team for use by the LSWR, sixty-six were built at Nine Elms and by Dübs and Co of Glasgow between 1899 and 1901. Robert Urie, post-1912, introduced a small number of modifications, which included the addition of superheating and an enlarged smokebox, to this well received class. No. 120, which is pictured above in its natural habitat, survived into preservation. (R. Hillier)

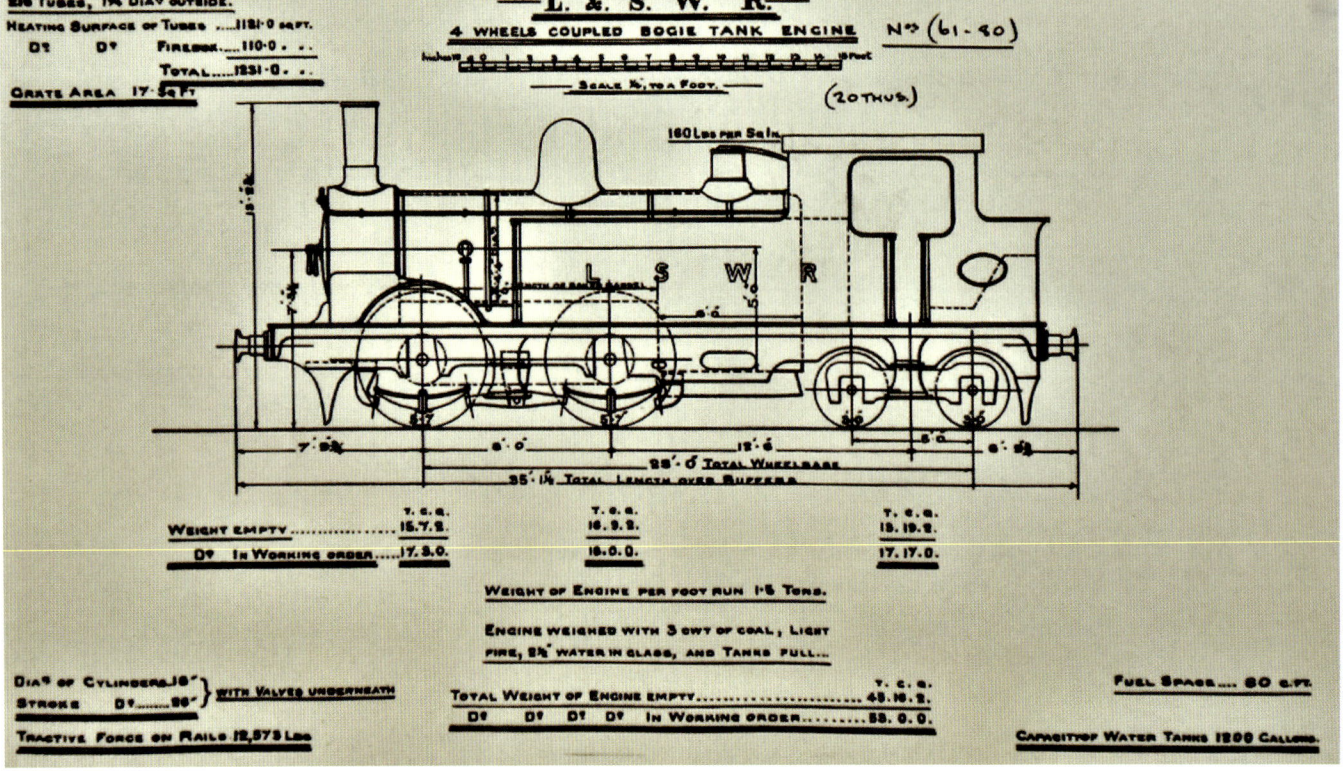

The T1 0-4-4 tank engine, six of which found service on the S&D during the war years, were a William Adams design for suburban passenger work. Fifty were built at Nine Elms between 1888 and 1896. Withdrawal of the class from service began in 1931 and but for the war they might all have disappeared sooner. As it was, the last few of them went in 1951. (R. Hillier)

freight and passenger trains undertaking essential war work. This could occasionally lead to 'foreign' engines traversing the line in addition to the locomotives more normally associated with its running. For the most part, the variety of engines used during these years has gone unrecorded. Luckily, Peter Smith, who for many years was a fireman then driver on the S&D, provided a few hints of what these years contained in his delightful book *Footplate Over the Mendips*.

He recalled the extent of the work that fell on the footplate crew and the many difficulties they, and all their comrades in the sheds or on stations, faced in keeping the railway running. In terms of locomotives, he recorded

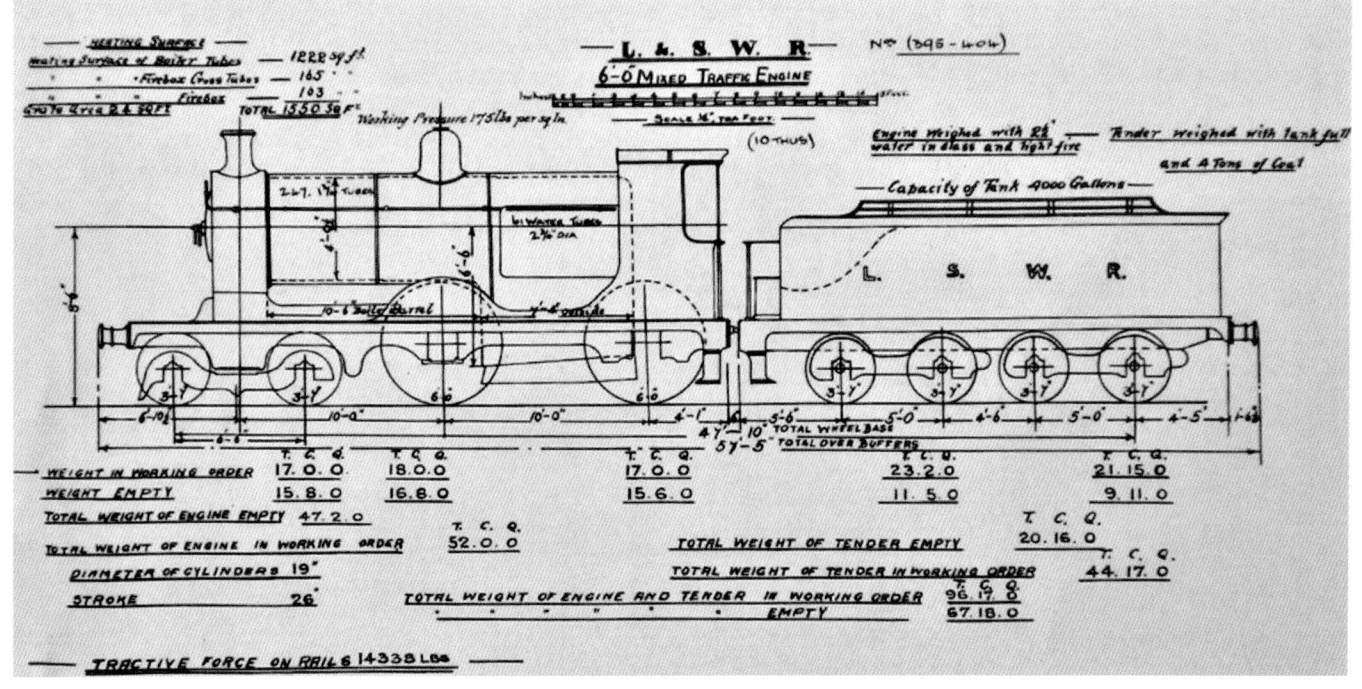

Peter Smith recorded that two S11/V11 Class 4-4-0s, Nos. 395 and 404, found use on the S&D but not how long they stayed. They were another Drummond design, derived from the T9s, with all ten of the class being built at Nine Elms in 1903. They survived in service until the 1950s, the last, No. 400, going in 1954. The locomotive featured in the photograph above is No. 396. (R. Hillier)

some of the comings and goings and the occasional 'rarities' that appeared from time to time. For 1941, he describes the departure of all but two, Nos. 5029 and 5440, of the S&D's Black Fives, 'much to the chagrin' of the crew. Three, 5023, 5194 and 5389, went to Perth, whilst 5289 and 5432 headed to Leeds, amidst the 'black-out and shortages of men and material'. It was a problem made slightly worse in 1943, when even the two remaining 5s were taken away, with Horwich Moguls, numbers unrecorded, arriving as substitutes. In describing their time on the S&D, he harked back to 1927 when engine No. 13064 underwent trials over the line and the disappointment felt when this didn't lead to a permanent allocation. Now they finally received some 2-6-0s, from Saltley and Burton sheds, including No. 2766 which 'was reliably timed to reach a maximum speed of 74mph at Sturminster Newton on an up express', gaining the crews admiration in the process for 'their sheer pulling power'. So they must have been pleased that they would remain on the line for some years to come, though 'when pushed hard it did tax their steaming capacity somewhat', unlike the Stanier 5s. However, they were less lucky with

the 8Fs which came for too brief a period before being recalled.

Nevertheless, there were some, mostly aged, substitutes of a kind in engines from the Southern Railway becoming available. These included some elderly T9 4-4-0s, No. 304 amongst them, a type not unknown on the line before and after the war, T1 Class Nos. 1 to 6, S11 Nos. 395 and 404, K10 4-4-0s for a short period early in the war. Then came an ex- South Eastern and Chatham Railway Stirling 4-4-0 No. 1188 and Drummond 700 Class 0-6-0s. However, these SR engines were not generally liked by the enginemen due to 'mechanical reasons' and the lack of space and refinement in the cab designs. They also found that the roofs didn't provide as good a protection as some Midland and all LMS locomotives that they had become used to.

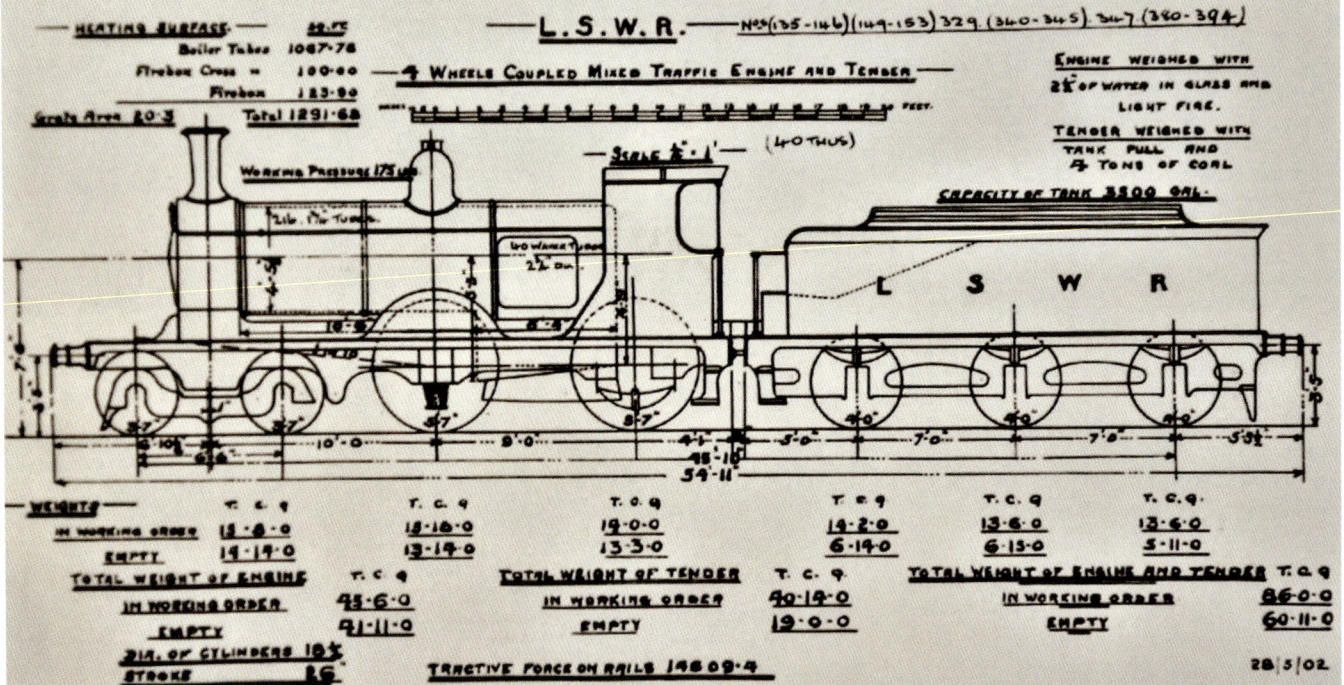

**The K10** Class were the last of the LSWR/SR 4-4-0 tender engines to be noted appearing on the S&D during the war. Designed by Drummond and team as a mixed traffic class the first of forty rolled out of Nine Elms in 1901 with production coming to an end in 1902. By 1940 they were thought to be unable to sustain power over long distances, so were deemed more suited to work over shorter secondary routes such as the S&D. The needs of war probably extended their lives with some remaining in service until 1951. (Author)

In this situation, the return of two Black Fives, Nos. 5056 and 4844 to Bath in late 1944, as the war in Europe reached its climax, must have been something of a blessing, more so post-1945, when demand across the network was returning to peacetime levels so allowing more of the class to be released for service on the S&D. This must have been a great relief to footplate crew struggling for so long with such a mixture of old engines. But, as Peter Smith later recalled, they did have the S&D 2-8-0s which 'contributed greatly to the war effort', with the '1914 series of these incomparable freight locomotive seeing out two world wars during their lifetime rendering yeoman service in both'. From a man with so much first-hand knowledge and experience of the type, there can be no higher accolade and in peacetime they

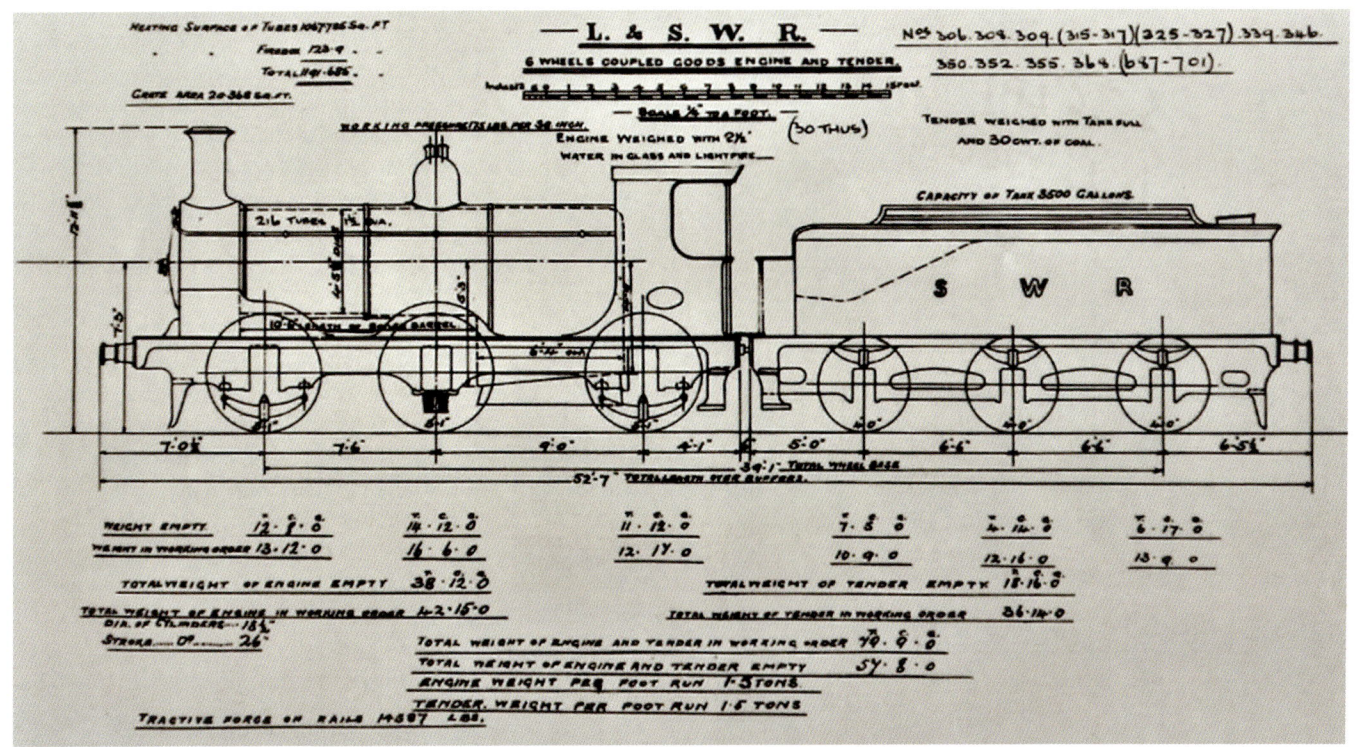

**Another Drummond** design, this time his 700 Class freight 0-6-0s, thirty of which appeared in 1897 built by Dübs and Co of Glasgow six years before they and two other companies, Sharp Stewart and Neilson Reid, came together to form the North British Locomotive Company. These engines contained many standard parts common to other Drummond locomotives, including the M7's boiler. The class underwent some modification during the course of its life, most notably when superheaters were fitted between 1919 and 1929. The success of the class might be gauged from their longevity, some remaining in service into the 1960s. Peter Smith recorded that these engines, nicknamed 'Black Motors', were mostly encountered 'on the Southern end of the line'. (Author)

continued to contribute greatly to life on the line.

Earlier in this chapter, Fowler's work in developing his 4P 4-4-0 compound locomotives was discussed. Although none of these engines were allocated to the S&D in peacetime, they were probably occasional visitors pulling expresses from the north. One, No. 1046, did become a fixture on the line for a time during the war. This 1924 Derby built engine came in for some rough handling when pulling passenger trains between Bath and Bournemouth, which it did quite often. Peter Smith surmised that this was caused by footplate crew more experienced in handling the 2P 4-4-0s forgetting that they were now on a compound, so a different driving technique was necessary. The tendency seems to have been to open the regulator fully when starting, which was fine for the 2Ps but not 1046 which went over to 'full compound working' when handled this way. The train would barely move and 'made some very distressed sounds and the train all but stalled'. The correct technique had to be learnt and then the crew began to make best use of the engine which proved superior as far as coal burnt per drawbar horsepower was concerned. Nevertheless, it lacked the ability to develop high tractive effort at low speeds on steep inclines so proved unsuitable for pulling the 230 ton trains for which it was intended.

A sad but essential part of military organisation is the need for rapid and effective treatment of casualties. The Great War had seen thousands wounded, at times on a daily basis, with facilities to evacuate them from the battlefield soon strained beyond endurance. Preparations for the Second World War proved little better, though the conflict at least didn't stagnate on one front and produce such an appalling daily haemorrhage of lives. Even so, the numbers leaving battlefields for treatment in Britain were still substantial and required the services of specialist ambulance trains to ship casualties from ports to hospitals across the country. The

**Seventy-one Class** S69s 4-6-0 were built between 1911 and 1928 for use by the Great Eastern Railway in pulling express trains from Liverpool Street Station in London to East Anglia. These Stephen Holden designed engines were taken over by the LNER in 1923 and redesignated B12, in which state they were gradually modified as this entry from Robert Thom's records reveal (Thom was the LNER's Mechanical Engineer based at Doncaster under Gresley, where he led in the construction or rebuilding of many engines, including the A4 Pacifics). With many new engines being built by this company, the B12s gradually slipped down the pecking order so were available and suitable to be assigned to pull ambulance trains. As such, they became a regular feature of life on the S&D during the war and for a time afterwards. (R. Thom/Author)

**LNER officials** and US Army representatives pose in front of a single carriage from a single ambulance train. Built by the LNER at Doncaster, they would soon be moving casualties from receiving ports to hospitals across Britain. These trains became a common sight on the S&D and a duty the footplate crew came to know well, especially with one based at Templecombe, two at Bournemouth West and the others at sites across the southern area. The B12 locomotives assigned to them were crewed for the most part by men from the LNER and Southern Railway with S&D men standing in at times, according to the few records that remain. (E. Thompson/D. Neal)

US Army sought the assistance of the LNER in creating ten of these trains using converted carriages and pulled for the most part by former Great Eastern Railway S69/B12/3 4-6-0 Class locomotives. They were chosen because they had wide route availability, were equipped with Westinghouse compressed air brake and vacuum ejector, making them suitable for pulling vacuum fitted stock. They were also available having largely been usurped on express passenger services by later Gresley engines.

During the war, these trains became a common sight on the S&D with local engines taking on piloting duties. For the most part, the B12s were crewed by LNER and Southern Railway men who lived on the train so as to be readily available. Inevitably there would have been times when staff numbers were stretched or depleted and S&D men had to act as stand in crew, especially during the twelve months that followed D Day in 1944. One train, given the number 36, was based at Templecombe, with its B12, No. 8549, shedded and maintained by local staff. Two others, numbers 27 and 31, found a home at Bournemouth West, from where they collected casualties from Portsmouth, Southampton and Weymouth mostly. This work continued into early 1946 as casualties from the Far East slowly returned home and the

**With the** war over, the railways could return to more normal schedules. They were suffering from overuse and reduced maintenance which had created a huge backlog of work. Each company also had to absorb the many surviving ex-employees who had left to fight the war. All this presented a very daunting set of problems, none more so than on the S&D. Although 7F No. 13806 presents a picture of normality after all the traumas of war as it hauls a coal train with a banking engine, the underlying message is one of change, though not necessarily for the better. (Author)

S&D saw an end to this essential but very sad work

Accidents both minor and serious will always be part of any industrial concern, but in wartime, when death and injury become horribly commonplace, day to day accidents lose their power to shock or cause close public scrutiny. Here the S&D were luckier than most, even though men and machines were worked to maximum capacity and, at times, beyond normal limits. Two accident books that survive show the usual bout of cut fingers, crushed limbs and cracked heads, some requiring hospitalisation. Interspersed with this were a small number of minor derailments of locos and rolling stock which seem to have been dealt with quickly without the need for a lengthy inquiry or major disciplinary action. But on 13 March 1944, this came to an end in a particularly horrifying and in a totally unexpected way.

With D Day so close the amount of military traffic along roads in the south of England was massive and growing ever larger. In such a situation, and with so many novice drivers working long hours day and night, accidents were inevitable, especially on local roads not designed for traffic of such size and volume. At Henstridge this proved to be a key factor in the accident.

The day began with a deep overnight frost and a sharply rising wind from the south-west, which failed to raise the temperature above freezing. As usual for 1944, traffic across the Mendips was heavy with a number of freight and troop trains, mostly heading towards the south coast. One of these was a ten-carriage troop train double-headed by LMS 4F No. 4523 and S11 No. 402, which led. Departing from Bath, conditions on the line, with a gale force wind hitting the train violently, proved very difficult and slowed progress considerably. At the same time, the crew on the 4F were finding it difficult to maintain steam pressure which made keeping to time even more difficult. Beyond Masbury, the two engines began to pick up time and were making good progress towards Henstridge, but the delay proved crucial in the events that followed. As they approached the bridge carrying the A30 over the track, the crew of the leading engine saw a heavy US Army transporter approaching, then appear to skid and hurtle towards the bridge parapet with such a force it could only hold for a brief second or two before giving way. Just as the two engines passed underneath, the trailer went over the edge. It struck the S11 and then fell between both engines, severing the connection as though a guillotine. This allowed the first engine to break away with pieces of wreckage dropping off before the crew could bring their engine to a halt.

The second engine then took the full force of the impact as the remains of the trailer and its load cascaded down over the 4F and its carriages, wreaking havoc as it went. In so doing, it severely damaged the engine's smokebox and caused it to derail and career into an adjoining field where it came to rest, followed by the first five carriages. The prime mover also toppled over the parapet but missed the debris of the train, although the driver was killed and the other occupant severely injured. Despite the seriousness of the accident, casualties were remarkably light – reported as one killed, five seriously injured and nine less so. Considering the number of men involved, this was something of a miracle. With Britain then operating under a strict code of secrecy, so that word of the coming invasion could not reach the enemy, coverage of the accident was muted. In addition, no official report by the Department of Transport seems to have been forthcoming and with this these events swiftly slipped from view only being revived much later through the memories of the S&D men who were there.

Such tragedies have to be endured with stoicism and courage, especially when so many people were dying each day around the world and in June the bloodletting would get much worse. With the allies restored to mainland Europe and making progress towards Germany, the end was at least in sight and with it thoughts turned to the makeup of post-war Britain with a degree of confidence that the war would soon end. The railways would cease to be under direct government control, though the slow handover following the Great War was likely to be repeated. Nationalisation had been on the agenda in 1918/19 and would be so again. But it needed a trigger for this to become reality and the Labour Party's landslide victory in the 1945 election proved to be the spark that ignited such a radical programme. It didn't happen overnight, but in 1947 the Transport Act was passed into law and nationalisation was enacted on 1 January 1948. The course of history was changed and with it came the eventual demise of the S&D.

**The S&D** under a new nationalised brand with BR numbers and tender totem. 7F No. 53809, having lost its larger boiler, rests between turns at Templecombe. (Author)

## Chapter 6
# A LONG GOODBYE (1948–1966)

**The new** corporate identity introduced by British Railways did not immediately lead to rapid change, except in the introduction of a new name and logo. With so many diverse companies to bring together, with their different working practices and design aims, the process of unification would take some time and face much opposition. The early years, even with complete collaboration and open minds, would have been difficult, but old allegiances died slowly and inhibited progress. However, a view of the future gradually formed and with it a policy of standardisation across the network, with steam locomotives still dominating and a nod given to the future by recognising the diesel and electric development work undertaken by the constituent companies. This painting produced in the early 1950s brought together these evolving ideas. Meanwhile the inherited fleets of engines rolled on with new production adding many more examples of existing types and new models. (Author)

**Nationalisation was** a new world to be explored by all its constituent parts, yet for some years it looked remarkably like the old world. Change when it came would be slow as BR tried to set right years of stagnation and damage caused by the war. But at least the communities served by the S&D no longer had to cope with the threat of bombing or the loss of its young men and women. Shortages and rationing continued for some years, but with the railway's chief competitor on the roads restricted by lack of fuel, they did enjoy a brief period of stability. Here, engine No. 13806 sits at Bath in the immediate post-war years still with its LMS number and presumably tender markings under all that grime. Unlike sister engines 13809 and 13810, this locomotive didn't receive its smaller G9AS boiler until 1955 and would remain in service until 1964. (Author)

The last eighteen years of the S&D's life were probably the most taxing of its existence. But they were also the most interesting for train spotters who were able to revel in the sheer variety of engines that appeared from across Britain's nationalised railway network. Some were allocated and many others simply visited, with few records now left to track the many types and their movements, particularly on busy Saturdays in summer.

Despite this simple pleasure for the bystander, these were years marked by uncertainty and fear of closure, despite the best efforts of its dedicated staff in trying to keep the line running and a going concern. The one bright spot lay in the way BR seemed prepared to supply locomotives that suited its needs very effectively and could have made a difference if continuation not termination had been the plan. The end result was probably never in doubt from the late 1950s onwards, during which time a backs to the wall fight was waged which attempted to wrest victory from probable defeat. However, in 1948 it wasn't clear in which way the future of Britain's railways would go and if cuts were avoidable or, if they weren't, how severe they might be. A key factor in this, the intensifying influence of the road vehicles on trade, had been checked by war and the recession that followed, exacerbated by the effects of rationing which lingered on into the early 1950s.

So in the new decade, the railway's biggest challenger lacked the teeth to compete, though this would soon change with lorries appearing in greater number, aided by the War Office's disposal of unwanted military vehicles, followed by more cars to meet growing social and business needs. In the meantime, industry still relied heavily on the railways to move its raw materials and the items it produced. Likewise, commuter needs remained high, supplemented during the summer months by ever-increasing holiday traffic as seaside towns shook off their wartime shadow and returned to normal. The S&D had never been a commuter line of any note, but it still had industry to support, particularly the coal mines. At the same time, holidaymakers from the north still continued to be a source of income for three or four months each year and remained so into the 1950s until car ownership boomed from about 2 million vehicles in 1945 to approximately 4 million ten years later. During the same period, heavy transport increased by another million or so vehicles. The railway's revenue across these transitional years may not have been high, but it was just sufficient to keep the wolf from the door. Nevertheless, it was based on the shifting sands of market forces and muted competition, which would soon change, aided and abetted by political and public pressures that favoured road over rail.

Against this background, nationalisation took root and its new management group, the British Transport Commission and its Railway Executive, soon found myriad problems to resolve, caused by wartime dereliction and lack of funds to correct many faults and problems. There was also the significant issue of bringing four strongly independent railway companies together as one. Few

**The changeover** from LMS to BR markings would take, in some cases, until the early 1950s to complete. Engine No. 53809 on show with the interim BR totem on its tender. Although this locomotive was consigned to the scrapyard at Barry in South Wales in June 1964 it survived into preservation and has worked regularly since restoration. (Author)

doubted that this might create insuperable problems and make unified working very difficult to achieve, with old enmities and rivalries difficult to shrug off. This proved to be the case. Nevertheless, a start had to be made somewhere and in due course new regional railways emerged – the Western, London Midland, Southern, North Eastern, Eastern and Scottish Regions – each with their own local management teams working to a single new centralised authority based in the Great Central Hotel at Marylebone.

The S&D, which had grown used to the LMS and the SR managing its business, now found itself in a new relationship. However, in some ways it was a play on an old theme as the line became the commercial responsibility of the Southern Region's District Controller based at Southampton. Under him, day to day control was exercised by the District Operating Superintendent of the London Midland Region at Bath, with the LMR retaining responsibility for motive power. It remained this way until 1950, when boundaries were redrawn and the Western Region took over the line north of Cole for commercial purposes, but this time with the SR managing the locomotives which were 'borrowed' from the LMR and the motive power depots. It was a move made permanent three years later when sixty-nine engines were formally transferred to the Southern.

Change once begun often becomes a permanent state and with greater economy and potential cuts still being the order of the day there was another boundary change in 1958. On 1 February, the WR took full control of the line from Bath to Henstridge and with it sixty-six locomotives and the sheds. In the process, BR decided to close the district offices at Bath. It was a programme of rationalisation that continued into the next decade. But in the first few years following the war this was a barely glimpsed future and a spirit of optimism still pervaded day to day life on the line, buoyed by a hard-earned peace.

**The three** men who would be responsible for shaping BR's locomotive fleet, and in so doing influencing life on the S&D. Here, in June 1948 they pose for this publicity photograph, with Cox pointing out some of the salient points in his report of future standardisation of steam locomotives. Left to right – Robert Riddles, who had been William Stanier's assistant and then led the Directorate of Equipment during the war. In this role, he led on producing austerity 2-8-0 and 2-10-0 class locomotives for the War Office. Ernest Cox, who became Development Assistant working for Tom Coleman at Derby in 1936 and then Chief Technical Assistant to Stanier and his successors during the war. In 1946, he was elected a member of the influential Association of Railway Locomotive Engineers so boosting his credentials for a senior role in BR. Roland Bond rose to become Works Superintendent at Crewe in 1941 and held this influential post until nationalisation. In 1953, he would become BR's Chief Mechanical Engineer when the Railway Executive was abolished. (E. Cox/R. Hillier)

In 1948, a key part of BR's planning focussed on the size and make up of its locomotive fleet. With 22,030 engines, made up of 448 different types, there was little standardisation or commonality of parts and this presented the Executive with a nightmare scenario when it came to choosing a sound economic path to follow. It was here that three men, Robert Riddles, Roland Bond and Ernest

Cox, all ex-LMS men, came to the fore, each chosen because of their known skill in managing a successful company at senior level. Riddles, as the Railway Executive's full-time member responsible for Mechanical and Electrical matters, was in all but name BR's first CME. Bond, meanwhile, was appointed Chief Officer for Locomotive Construction and Maintenance and Cox Executive Officer for Design.

Cox later described the problems they faced in trying to bring many threads together in a plan for the future. He wrote in a letter:

'One thing was impossible, namely to satisfy all interested parties to whom the Technical aspect now became overcast by political thinking. We could simply have halted all steam locomotive development, continuing with existing regional railway designs until large scale electrification and diesel schemes matured. But the state of design work on these new systems and the cost was such that it would take many years for such a changeover to take effect, so a more immediate solution was needed while these plans reached fruition.

'We could have accepted a continuation of steam and let each region develop new locomotive types to meet their own needs. However, one key remit from the BTC was to realise a high degree of standardisation, which such a solution would not have achieved. So, we could have wiped the board clean and designed a completely new series of standard locomotives covering the whole power range owing nothing in concept or detail to what had gone before. Then again we could have followed what Gresley did following amalgamation in 1923 and designed a small number of new types centrally for general use, but continued to build the designs of the constituent companies for all other duties.

'In the outcome we did none of these things exactly and instead turned to fact finding. Riddles at once initiated this

**If the** casual observer had chosen to visit the S&D's sheds at Bath in the pre-war or immediate post-war years, they would have been hard pushed to see any difference in the locomotives they saw. Even with nationalisation, the engine numbers and tender logos still reflected the LMS influence, as witnessed by these two Class 2P 4-4-0s, Nos.696 and 697, which were both built at Crewe during 1932. Neither locomotive was based permanently at Bath, so would appear to be temporary additions (697 carries a Bristol Barrow 22A plate, the other one is too grimy to see). Both of these locomotives survived until 1962. (Author)

process, four measures of which concerned locomotive design – fittings and renewable parts, the performance of regional locomotive types, locomotive testing and interchange trials. Lastly, I was asked to prepare a report summarizing all these findings and attempt to define a series of twelve standard locomotive designs. This became the blueprint upon which the ultimate design of the standard engines was based.

'Two of the twelve types suggested – a Class 8P four-cylinder 4-6-2 and a two-cylinder 2-8-2 – were laid aside for lack of immediate traffic demand and when they came to be built the former had three-cylinders and the latter had been transformed into a 2-10-0. The rest were built largely as proposed. To do this each of the design offices at Derby, Swindon, Doncaster and Brighton were given a dual task. Each was made a parent for a particular type or types and to this was added responsibility for designing a range of standard components that could be used across the range of new locomotives.'

The end result of this work was the eventual construction of a standard range of twelve classes of steam locomotives between 1951 and 1960, eventually totalling 999 in number. At the same time, the regions were allowed to continue building their own established designs and in the period 1948 to 1956 this added another 1,538 engines to BR's fleet, offset to a certain extent by disposals of older engines.

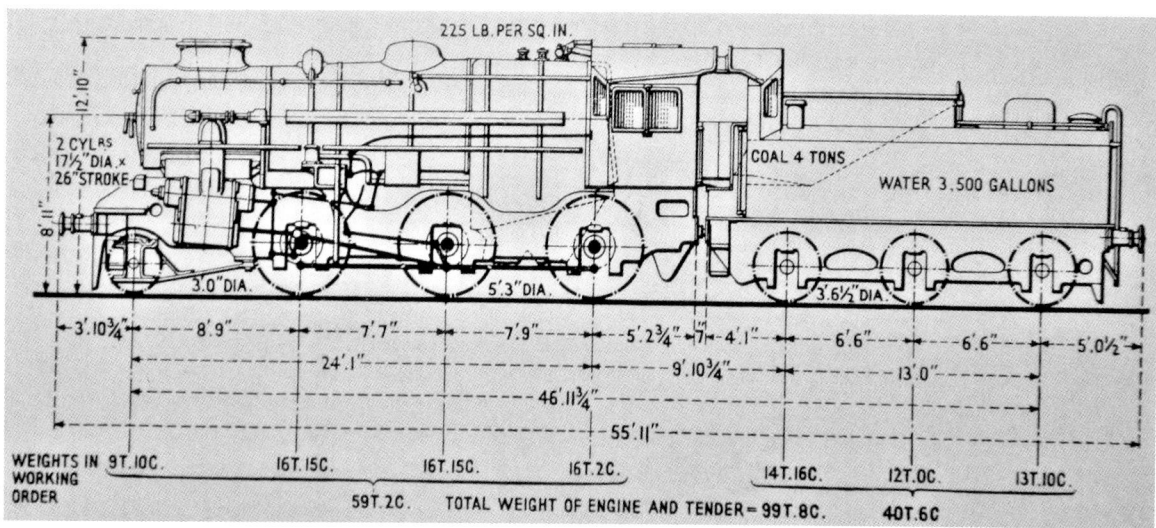

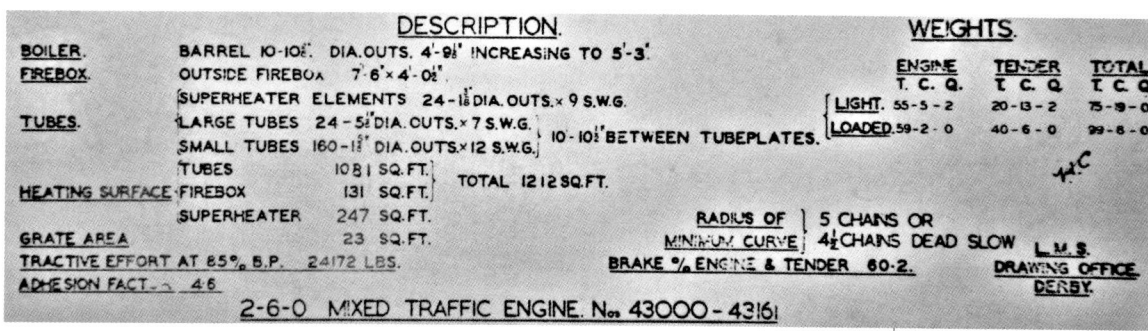

**In 1946/47,** with Henry George Ivatt now as the LMS's CME, locomotive production included two mixed traffic 2-6-0 engines designated 2MT and 4MT. A single 4MT found its way to Bath in 1948 and worked during the summer season over the S&D. In the years that followed, this engine returned, as did three other members of the class. As built, they were fitted with a double chimney but this with its accompanying blastpipe arrangement were found to perform poorly and in due course a single chimney was fitted. Production ran from 1947 to 1949, by which time 162 had been built. (Author)

It was neither modernisation or standardisation, but the product of an unmanageable situation exacerbated by the country's poor financial state and an addiction to coal which still employed more than 700,000 people in its extraction. Dieselisation, or electrification for that matter, made more practical sense, but other pressures and lack of political will made these impossible to achieve for many years to come. However, if given greater encouragement to develop more quickly, would internal combustion locomotives have saved the S&D? Probably not, because the indications are that the line was deliberately allowed to wither, but in a fairer world who is to say whether diesel engines might have made it more economical to run and made the decision to close a more difficult one to take? As it was, there was no flood of new diesels to report and new standard steam engines arrived instead to play a part in the S&D's battle for survival.

Until these new engines began to appear, the line carried on pretty much as it had done in the pre-war years. However, there were some changes to the its infrastructure to be absorbed, perhaps hinting at a wider programme of cuts and efficiency savings to come. Templecombe Lower Yard closed in May 1950 and the Wells branch line followed suit in October the following year. At the same time, passenger services between Highbridge and Burnham were terminated, except for occasional through workings by excursion trains. In 1952, the Bridgwater Branch was also closed to passenger trains.

Engine types didn't vary a great deal in these early years whilst the ravages of war were slowly put right, and each locomotive underwent periodic repair and maintenance. However, there was one noteworthy change in the summer of 1948 when a newly built Ivatt 4MT, No. 43102, was temporarily allocated to the line to help with increasing volumes of passenger traffic from the north. It proved to have poor steaming qualities, linked to its double chimney, and this meant that it tended to pilot other locomotives rather than work main line services by itself. Nevertheless, with 162 Horwich built engines eventually in service it was probably unavoidable that this locomotive, plus three sisters, Nos. 43013, 43017 and 43036, would become fixtures on the line, no matter how poor they were. However, there was some improvement when a single chimney was fitted and tests undertaken at Swindon resulted in modifications being made to their draughting that allowed the boilers to achieve something nearer their full potential.

Another product of the Ivatt/Coleman partnership, which lasted until 1949 when the Chief Draughtsman retired, arrived shortly after the 4MTs, and these engines seem to have received a warmer welcome. The first nine Class 2 2-6-2 tank engines, which

**Ivatt 4MT** No. 43017, the third engine of this class to be allocated to the S&D. Here the engine is photographed on the Midland side of Birmingham's New Street Station as it acts as 'station pilot' pulling empty stock. This engine appears to have been working from Bath in 1949 but was then assigned to Saltley the same year. However, it was back in the West Country during 1953 where it remained until 1964. By the time this photograph was taken, the double chimney had been replaced by a single version. (Rail Online)

were built at Crewe, made their appearance in 1946. By 1952, another 121 had been added to their number, the final ten from the workshops at Derby. In many respects they were the twin of Ivatt's Class 2 2-6-0s which appeared in the same year. Each shared a common design philosophy, were straightforward in concept, contained many standard components and were easy to maintain and operate. Both were designed at Derby by Coleman and a greatly depleted drawing office team during the war when Charles Fairburn was CME but had to wait for the end of the conflict for construction to begin.

The requirement for the tank locomotive was for a small engine that could undertake light, mixed traffic duties and have wide route availability. In service, it proved successful in this role and during the latter part of 1949, two, Nos. 41240 and 41243, were allocated to Bath, with a third, No. 41241, being added early in 1950, to be based at Templecombe. The few reports that remain suggest they were 'good, comfortable runners' had excellent steaming qualities, were very effective in service and were well liked by footplate crew and workshop staff equally. Like the 2-6-0 4MTs, they underwent testing at Swindon, which identified improvements that could be made to their draughting arrangements, and this enhanced their performance still further. These engines proved so valuable

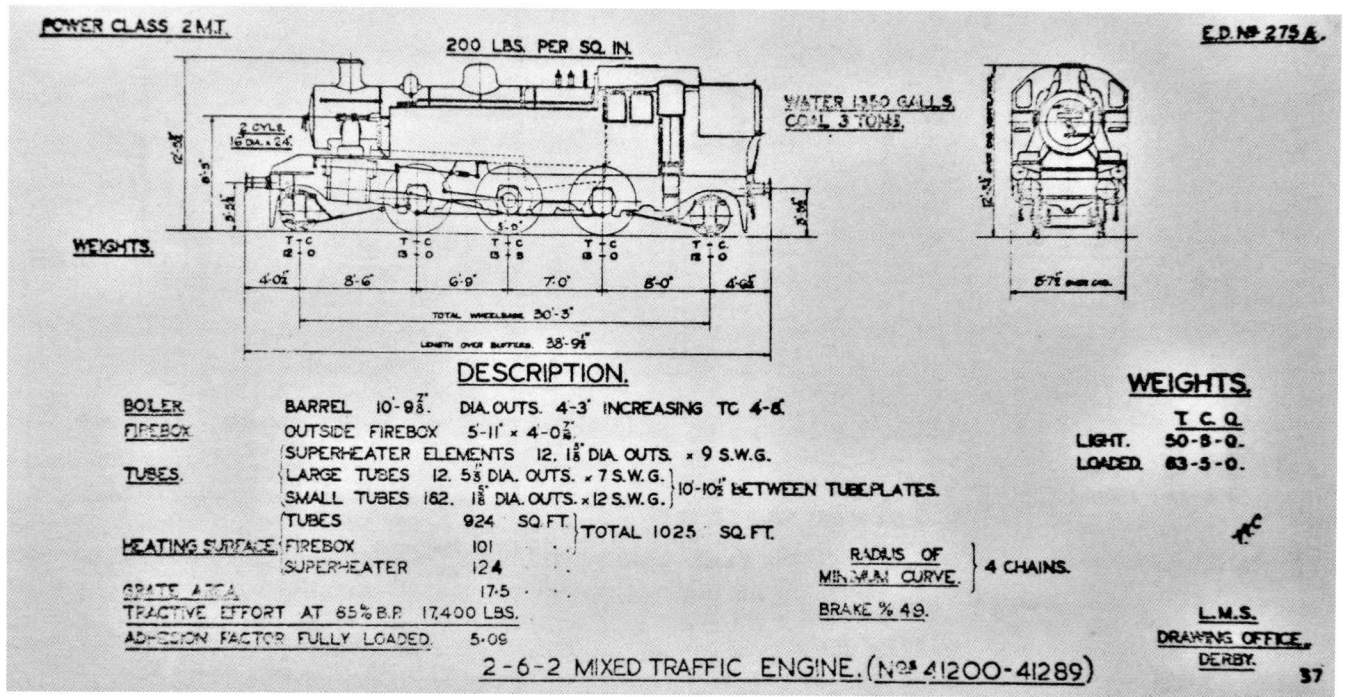

**Tom Coleman** later described the Class 2MT 2-6-2 tank engine that first appeared in 1946 as 'a good all round design that met a specific need very effectively and found a second wind when modified by BR into the standard Class 2 2-6-2T introduced in 1953'. The diagram to the left bears his initials and the photograph shows the first engine of the class painted in primer grey at Crewe to be photographed for publicity purposes.
(D. Neal)

to the S&D that samples of the class remained on the line until its closure.

The post-war years for the railways proved to be a very difficult time in a number of ways. It is certain that the poor state of the network, following years of debilitating work and lack of investment, had left the infrastructure sorely in need of major work and modification to repair the damage and bring it up to a reasonable level. Safety was undoubtedly compromised and the number of accidents, many very serious, probably reflected the parlous state of the system. Major disasters at Lichfield, Bourne End, Balby Bridge, Sutton Coldfield, Winsford, South Croydon and Harrow and Wealdstone made the headlines, but there were a substantial number of other incidents to record. And yet many of these were simply accidents that might have happened on even the best regulated and modern networks. A misty morning on 26 August 1949 proved just such a case.

Ground hugging early morning mists on the Somerset Levels during the hot and humid summer months are not uncommon. Along South Drain, which runs for nearly 9½ miles from the Actis Tunnel near Glastonbury towards the Street Area before turning north-west towards Ashcott, where waters from the Glastonbury Canal enter, this is especially so. Poor visibility, and possibly a poor lookout by the crew of the 8.5am mixed passenger and goods train, hauled by 3F 0-6-0 No. 3260, was a contributory factor in an accident to the west of Ashcott station. Here a 2ft gauge line from the Eclipse Peat Company's factory crosses over the S&D's line to reach the peat beds nearby. With time to spare in the timetable and good lookouts, this would have been normal day to day business, but on the 26th, this engine stalled and couldn't be re-started. With a train due, the driver ran towards the approaching locomotive, but was unable, due to the mist and possibly the crew's lack of attention, to bring it to a halt. Although the peat engine was small and offered little resistance it was sufficient, with its trucks, to derail the 3F which had enough momentum to carry on for a short distance before slipping down into South Drain. As it did so the driver and fireman leapt for safety, while the carriage and trucks remained upright on the track above. Salvage in such conditions proved very difficult and the water course had to be diverted around the engine to allow it to happen. But this could only be achieved by laying a bed of railway sleepers over which a makeshift crane could pass and then cutting up the engine into manageable pieces for scrap. So ended the life of one of the S&D's longest serving engines.

Although BR had yet to embrace dieselisation in a meaningful way, there had been some progress before the war and in its immediate aftermath. In particular, the GWR had experimented with railcars and the LMS with its two main line engines Nos. 10000 and 10001. But it was the first of these that had an impact, albeit a small one at least, on the S&D. In the early 1930s, with Charles Collett as CME, a need for small, mobile units to serve the company's secondary passenger routes was identified. Years earlier, Churchward had considered diesel railcars as one solution to this, but no strong, reliable internal combustion engine was available. All this changed when the Associated Equipment Company (AE. Cox) of Southall developed its six-cylinder engine which produced 121hp at 2,000rpm. These were installed in their standard Regal buses with much success. Seeing the potential for its use on the railways, AE. Cox produced a rail mounted version with flanged wheels which it offered to the GWR. Collett supported this proposal, but insisted on a heavily modified type to meet the more demanding needs of railway operation.

Design work by C.F. Cleaver of AE. Cox proceeded with the Hardy Railcar Co, an AE. Cox subsidiary, taking the lead in construction. In 1933, the first car emerged, most notably in streamlined form and capable of carrying sixty-nine passengers, to enter service on the lines from West London into Berkshire. Such was their popularity that construction carried on, with various modifications, until 1942 by which time thirty-eight were in service, two of them as parcel cars. Some of these vehicles, five at one time, were based in Bristol from where they occasionally found their way onto the S&D. For the most part, they seem to have trundled between Temple Meads Station and Bath Green Park, via Mangotsfield, offering an alternative service to that from the Bristol to Bath Spa Station line.

There are some vague reports suggesting that they may have ventured further south along the S&D's line, but nothing to say that it was anything more than a temporary expedient to cover

(Top) 3F Class engine No. 3260, built by Neilson Reid in 1902, part submerged in the mud and water of South Drain that ran beside the line from Glastonbury, on 26 August 1949 following a collision with a small petrol engine to the west of Ashcott. The locomotive helping clear up is another Neilson 0-6-0, in this case No. 3248. The second photograph is believed to show an Eclipse Peat Company's light engine of the sort that stalled on the line that morning as it pulled two trucks. (D. Neal)

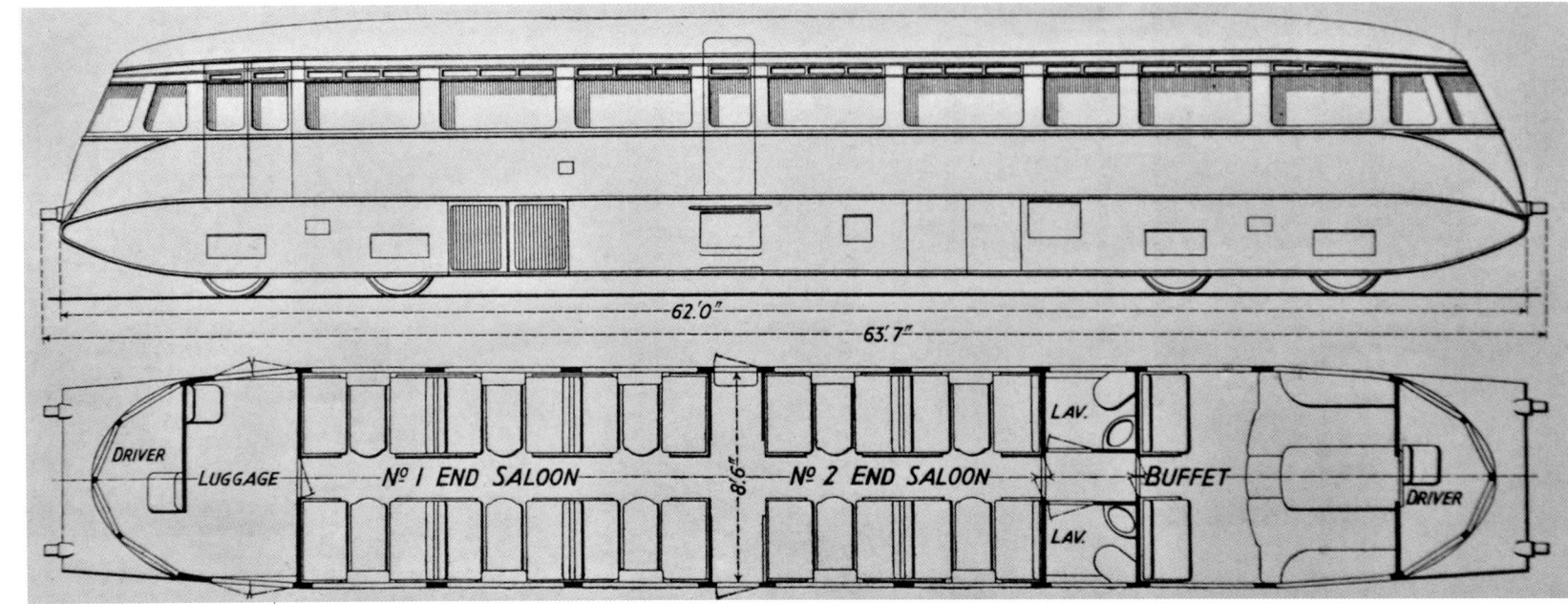

**After the** prototype railcar entered service, a methodical rather than dynamic production programme got underway. Three more of the type illustrated here were built in 1934, followed by another three in 1935, ten in 1936/37, ten in 1940 and a final eight in 1941, these two final groups being erected at Swindon. They saw wide use across the GWR system and found ready employment with BR with a number remaining in service until the early 1960s. (Author)

a shortfall in motive power. Ivo Peters, in his many expeditions as he recorded the company's everyday life, occasionally photographed these railcars in action suggesting their appearance was not unusual between Green Park and Bristol at least. There is also a hint, in a 1941 article in the railway press, that a single GWR railcar was in the process of being tested over the line but gives no other details or confirms the results of this work. However, as no allocation followed it may be assumed that little was achieved by such a wartime experiment. Nevertheless, these reports, if they are to be believed, do suggest an interesting possibility that some dieselisation might have been possible years before the S&D's closure became a realistic possibility. If so, what might have been achieved in terms of increasing local passenger traffic and revenue? Possibly not a great deal if truth be told. Undoubtedly, they may have been more cost effective to run, especially as the price of diesel fuel gradually made it a viable alternative to coal. Also, just turning up and switching on an engine to start was a lot quicker than the prolonged process of preparing a steam locomotive for service. The later railcar models were also equipped to work in tandem so offered a degree of flexibility in meeting fluctuating traffic demands. There was also the question of dependability and maintenance to consider. With diesel engines growing in reliability as the technology improved, railcars were found to run an average of 300 miles per day without problem. And with their lives extending over a 20 to 25-year period they seem to have survived fairly well, with the arrival of more advanced multiple units being the primary reason for their demise. Despite all this, it is clear that the line's problems stemmed from the sparsely populated, largely rural community it served, and not whether steam or diesel were employed, no matter how cheap or effective they may have been. But the S&D weren't alone in facing this problem, across Britain the same reasoning applied, and closure beckoned for many lines.

The railcars weren't the only diesel powered passenger trains to offer the S&D a passage into a future beyond steam for its passenger traffic. By the early-1950s, BR had begun building a number of multiple units that would gradually populate many lines across the network. The first were two car sets, twenty-one in number, built at Derby specifically for use in the West Riding of Yorkshire and West Cumberland. The first sixteen power cars were fitted with two Leyland six-cylinder engines and the later ones with AE. Cox 150hp units. Such was their

**Railcar No.** W24W was a regular, running between Bristol and Green Park and is captured here in its crimson lake and cream livery approaching Temple Meads Station in the early 1950s. Could these railcars have worked over the S&D and help boost its local passenger traffic? Probably not because the line's problems ran more deeply than this, but it is an intriguing possibility. (Rail Online)

success that the programme rapidly expanded and BR's Modernisation Plan, published in December 1954, included a requirement for 4,600 diesel multiple units to aid the replacement of steam locomotives. A significant ambition but one unlikely to be achieved quickly. However, by 1958, three, four and eight car sets were in service and two of these briefly visited the S&D in May that year to be recorded by Ivo Peters in his travels. However, it wasn't part of a formal trials process but simply involved day excursions. On 10 May, a three car set with a group of enthusiasts on board, organised by the Gloucester Railway Society, traversed the line and on Whit Sunday a Birmingham to Bournemouth eight car set was in evidence. By this stage, the Western Region had taken over control of the line as far south as Henstridge, so it seems likely that run down and closure were already on the agenda. In this event providing more cost-effective solutions, such as DMUs, wouldn't have rated very highly in the regional managers' thoughts. Despite this, these units continued to make odd appearances on the S&D, fulfilling excursion duties and appearing to meet the demands of the line, with its steep inclines, fairly well.

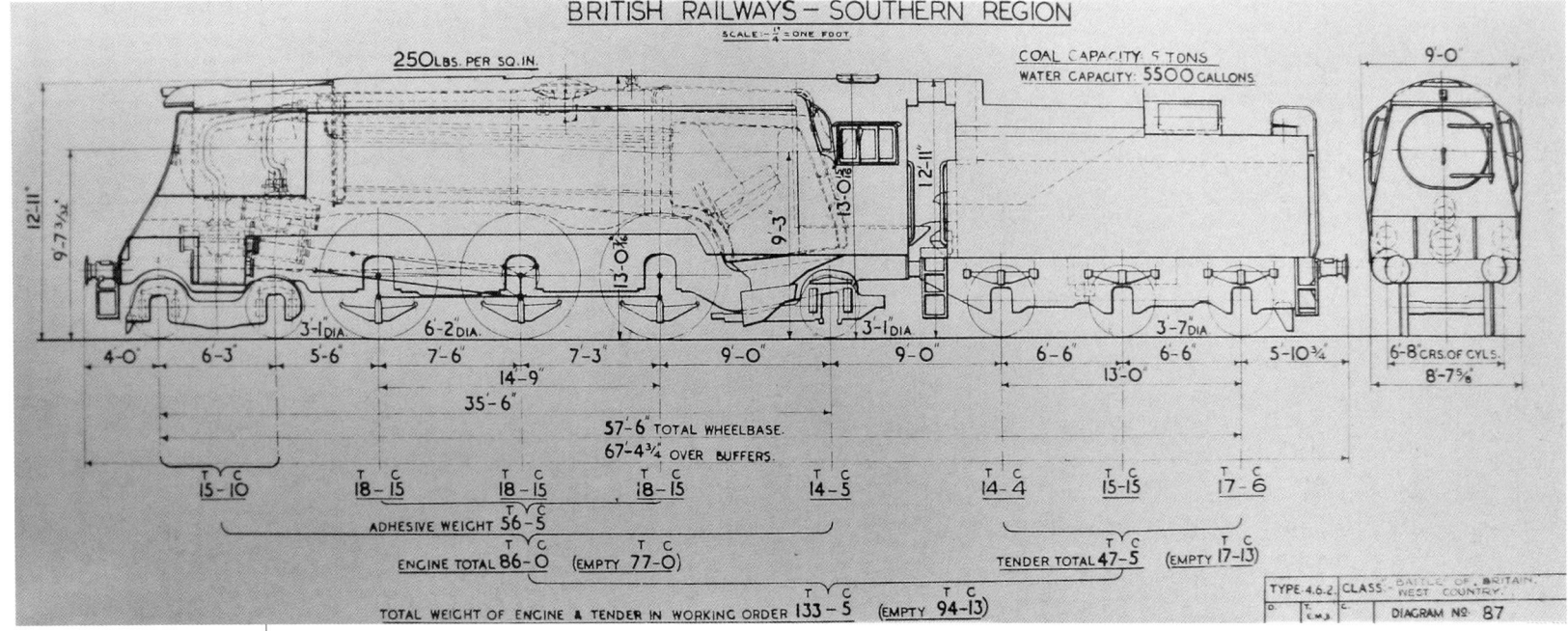

**Oliver Bulleid's** Pacifics, both light and heavy versions, had a Marmite quality about them. People seemed to love or hate them in equal measure. There would also be a division of opinion when the majority of them were later rebuilt, between supporters of both versions. In truth, they were good engines but Bulleid's fertile mind and desire to experiment may have been their undoing. Several drivers later recalled in their memoirs that they steamed better in their original condition, but as rebuilt were much more reliable 'but lacked the swagger of the originals'. In 1951, footplate crew on the S&D were given a chance to find out for themselves. (Author)

The Southern Region's influence over the S&D, which continued to grow until 1958, meant that some of its engines found their way on to the line to replace LM classes. There had always been occasional visitors, of course, but now more permanent arrangements were made, especially in the summer months when through passenger traffic reached a peak. By 1951, the Southern had 30 Merchant Navy and 110 West Country/Battle of Britain Class Pacifics in service. There were probably more than were needed, especially as they were failing to live up to all expectations, displaying some interesting characteristics in the process. Although regarded as powerful and free steaming, they presented a number of challenges to footplate crew and maintenance staff alike. Their coal consumption was heavy by comparison to other Pacifics, the valve gear often failed in service, due to fractured rocker shafts and driving chains, which also tended to sag when in constant use, and their oil baths frequently leaked and were subject to water ingress causing corrosion of the motion parts. Fires were a frequent occurrence, thirty-eight recorded in 1953 alone, caused by 'accumulations of oil-soaked inflammable matter' gathering around the ashpan hopper doors. These fires often spread to the boiler lagging which itself had been soaked by vapour or oil leaking from the oil bath.

When in service, these engines had a propensity to slip violently, making them difficult to handle at times and a crash at Crewkerne revealed a weakness in their middle driving axle. 35020 *Bibby Line*, when pulling a Plymouth to Waterloo train on 24 April 1953, suffered a catastrophic failure of the crank axle. By good luck, the train remained upright, though demolished part of the station, and there were no major injuries. Investigation by a supersonic flaw detector revealed a significant problem that was likely to affect all the SR Pacifics, both heavy and light. All in all, the negatives were seen to outweigh the positives and in 1955, BR put in hand a rebuilding programme that would eventually deal with all the Merchant Navy Class and sixty of their smaller sisters before steam disappeared from the Southern in 1967. It was against this background that the light Pacifics found their way to the S&D, though not the 'Packets', as they were nicknamed, because their axle loadings were too high.

The first Pacific to arrive was No. 34109, *Sir Trafford Leigh-Mallory*, and it came on trial during March 1951. This engine was only ten months old so was in prime condition. Before its first run from

**While pausing** to take water at Evercreech Junction Station, West Country Class Pacific No. 34042 *Dorchester* receives some attention from the crew. The date isn't recorded but it will be before 1958 when this locomotive was rebuilt. 34042 was allocated to Bath twice, both times in 1951, May and then December, remaining in Somerset until transferred to Eastleigh during November 1954. When rebuilt it was allocated to Bournemouth and would again appear on the S&D occasionally. In September 1964 this locomotive transferred back to Eastleigh and was withdrawn in January 1966. (D. Neal)

Bournemouth to Bath on 14 March, it was prepared at Branksome by men unfamiliar with the type, with the result that they loaded too much coal into the firebox making combustion difficult. However, with a load consisting of only four carriages, the demand placed on the engine was not excessive and it coped. However, it was a problem that re-occurred in these early days, on one occasion causing the engine to fail at Templecombe and be substituted by a 4F. It was only a matter of experience, though and the firemen soon learnt the correct technique and with this began to coax much better performances from this engine with trains upwards of ten carriages to Bournemouth and back.

With so many light Pacifics available, others soon arrived; 34109 plus three others being allocated to Bath, with others borrowed from Bournemouth Shed as required. By the end of the 1950s, at least thirteen had seen use on express services over the line and the next decade would see this continue for a time with rebuilt engines added to the list. However, by 1962, they had largely been superseded by the exceptional 9F 2-10-0s, the first of which ran trials over the line in 1960. The Pacifics still appeared after that but spasmodically and never returned in the same number again. In one final flourish of activity, two immaculately turned out light Pacifics, 34006 *Bude* and 34057 *Biggin Hill*, pulled a Farewell to Steam special on 5 March 1966 to Bath and back to Evercreech in celebration of the line's last day of operation. Shortly before this, two Merchant Navy Class engines, 35011 and 35023, had been allowed over parts of the S&D. They were chosen to pull special trains, their axle loading being ignored now that the track was soon to be lifted.

The Pacifics weren't the only Southern engines to make an appearance in this last phase of the S&D's life. In March 1954 Maunsell U and U1 Class 2-6-0s each spent

**A busy** Saturday in the 1950s sees an unidentified West Country Pacific pulling a rake of crimson and cream carriages over the Mendips with 2P No. 40568 acting as pilot. This became a common sight which didn't change appreciably when the Western Region/Southern Region boundary changed again in 1958. These Pacifics continued serving the S&D until the mid-1960s. (D. Neal)

a week undergoing tests to see how successful they might be in accomplishing everyday tasks. Working from Bath, engines 31621 and 31906 undertook a variety of tasks, but didn't impress sufficiently to justify the allocation being made permanent. As Ivo Peters observed, in volume one of his four book photographic history of the line, the Schools Class 4-4-0, No.30932, that worked an enthusiasts special to Bath a month later represented a type that would probably have suited the S&D much better than the U Class.

As the Southern Region sought to introduce its engines to the S&D, the products of BR's standardisation plan were reaching fruition. In 1954, they began appearing in Somerset.

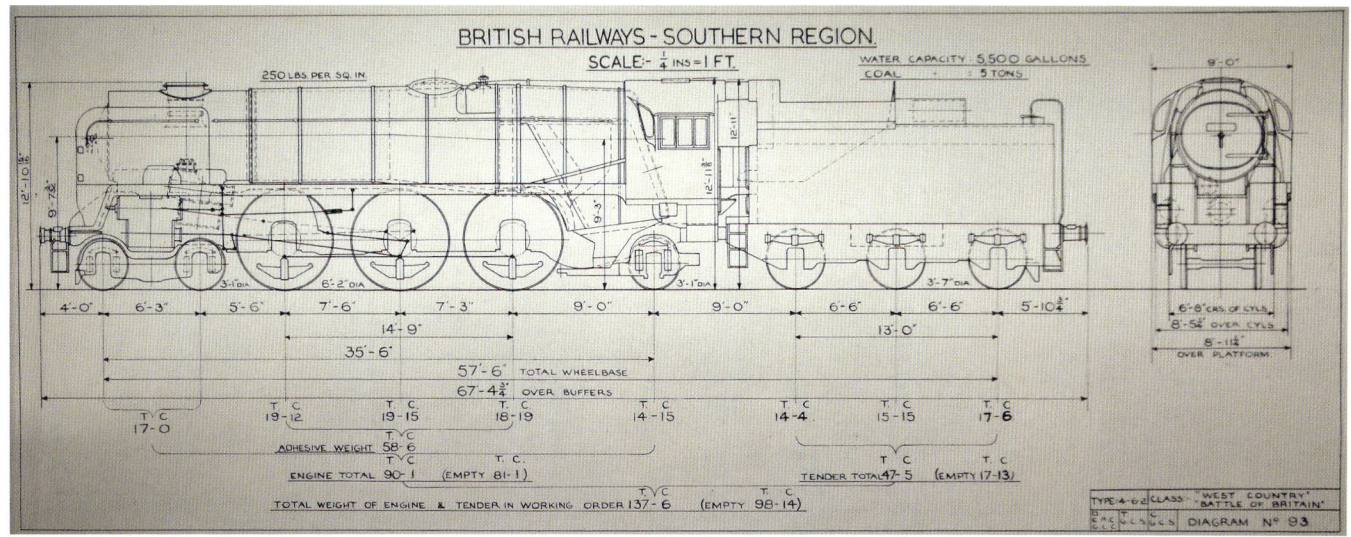

**The rebuilding** of many of Bulleid's Pacifics rested in the hands of Ron Jarvis, the Southern Region's gifted Chief Technical Assistant at Brighton, who worked directly to the Chief Mechanical and Electrical Engineer. He had the vision to take the best of Bulleid's work and bring a strong practical element to the design, helping to eradicate some of their more troublesome features. At the same time, he made improvements to some of the originals which seem to have improved reliability. Whatever the strengths or weaknesses of either version, each type continued to serve BR until steam vanished from the Southern in 1967. Top – The rebuilt engines, apart from the Bulleid Firth Brown 'boxpok' disc wheels, almost unrecognisable from the originals. Below – 34040 *Crewkerne*, one of the few Pacifics to appear on the S&D in both guises. Constructed in 1946 and rebuilt during 1960, it spent the last years of its life based at Bournemouth from where it would occasionally venture on to the S&D. This engine was withdrawn in July 1967 and scrapped in March the following year by Cashmores in South Wales. (Author)

**Formal trials** were held with Richard Maunsell's U/U1 Classes in 1954 to gauge their suitability for use on the S&D. They didn't find favour and the idea was quietly dropped. However, these engines occasionally visited the line as shown here with U Class engine No. 31795, pulling a day excursion service near Templecombe on 23 July 1956. The photo was taken by Peter Pike who was then a fitter at the sheds nearby. (P. Pike)

**Other occasional** Southern Region visitors to the S&D were the 0-8-0 Z Class shunting engines. Eight were built in 1929 for use in marshalling yards, but during the 1950s, this role was taken over by diesel electric shunting locomotives. As a result, the entire class were transferred to the SR's Western Section where they found employment undertaking banking duties, most notably in Exeter. However, as this photograph reveals, engine No. 30953 found employment at Templecombe on shunting duties in the Upper Yard and occasionally as station pilot. (Author)

**Although the** arrival of Bulleid's Pacifics was supposed to presage the departure of Stanier's Black Fives, as did the production of BR Class 4 and 5 4-6-0 standard engines, they still appeared on the line quite regularly pulling trains from the north post-1958 until the early 1960s. Most were basic 'Fives', but a small number of these engines, fitted with the British version of the Caprotti valve gear, also ran over the line. This picture captures engine No. 44827, built at Derby in 1945, seeming to make light work of its load in the late 1950s. On 26 June 1962, Ivo Peters recorded the presence of Saltley based engine 44841 double-heading the down Pines Express with 9F No. 92233. This may be the last occasion on which a Black Five worked over the S&D, but another Stanier stalwart, the 8F, would return to the line after many years away to replace the long serving 7Fs as they were withdrawn from service. (R. Hillier)

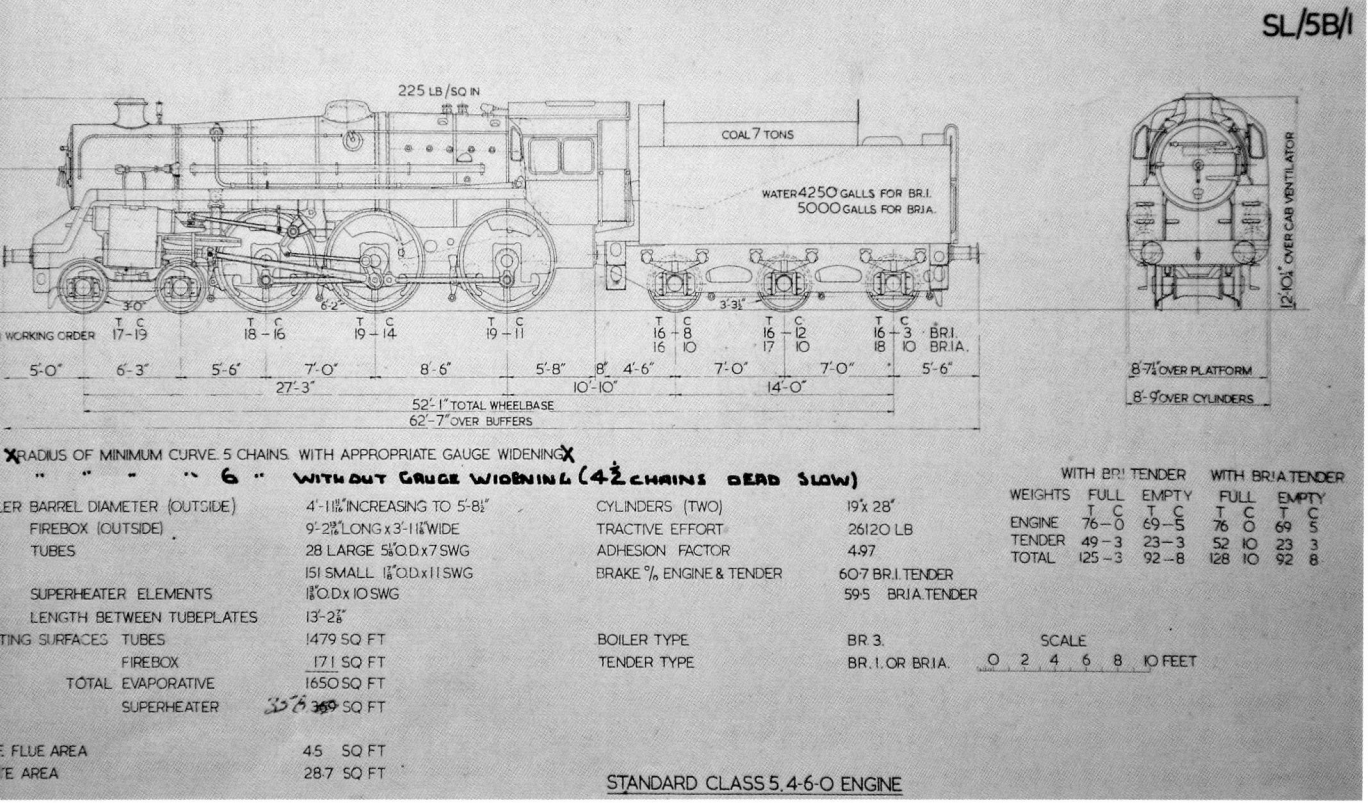

**Doncaster drawing** office led in creating BR's Class 5 4-6-0 mixed traffic engines and produced a well-balanced design that proved to be strong and reliable. In due course, 172 were built and found wide use across Britain's railway network including the S&D with three engines arriving in 1954. By the time the line closed, at least twelve of the class had worked from Bath, either as part of a permanent allocation or simply passing through. (Author)

**A sight** that became very familiar on the S&D in the 1950s and 1960s; a Standard Class 5, on this occasion engine No. 73082, built at Derby in 1955, photographed when about to depart from Evercreech Junction Station in the late 1950s. (Author)

The programme commenced with the arrival of three Class 5 4-6-0s, Nos. 73050, 73051 and 73052, the first of these on 19 June having been present at the International Railway Congress Association Exhibition at Willesden. Their arrival didn't come a moment too soon. The unsuccessful tests involving the U and U1s had reduced the SR's options for adding extra locomotives to the S&D's fleet for the busy summer months. So it became necessary to find a viable alternative and the Class 5s fitted the bill. All three locomotives allocated were built in 1954 at Derby, No.73050 in April, the other two a month later, as part of a programme that began in 1951 and ran on until 1957, by which time 172 had been constructed.

When Ernest Cox first considered the range of engines to be included in his draft standardisation proposals, he envisaged the need for a mixed traffic Class 5 power classification being met by a Pacific. As time went on the 4-6-0 wheel configuration replaced the 4-6-2, a decision possibly influenced by the highly successful LMS Black Fives and LNER B1s, let alone the many other 4-6-0 engines that had emanated from Swindon. The task of designing the class fell to the drawing office at Doncaster, under its highly experienced Chief Draughtsman Edward Windle. In due course, a second standard engine, the Class 4 2-6-0, was added to his brief, plus a number of components to be fitted to other new classes - cylinders, slidebars, crossheads, coupling and connecting rods and valve gear. Windle, who served the three LNER CMEs, Gresley, Edward Thompson and Arthur Peppercorn, contributed hugely to many successful designs including the A4 Pacifics and B1s, and was a designer of some substance and experience. As such, there was probably no better man to lead this work and the end result was deemed to be a success, with the class proving to be strong, reliable and possibly equal to larger Pacifics in service.

The next standard class engine to appear on the S&D was the Doncaster designed Class 4 2-6-0, but on this occasion, formal allocation was preceded by a brief trial. On 5 March 1955, engine No. 76012 pulled the 12.55pm stopping service from Bournemouth to Bath, piloted by 2P No. 40568,

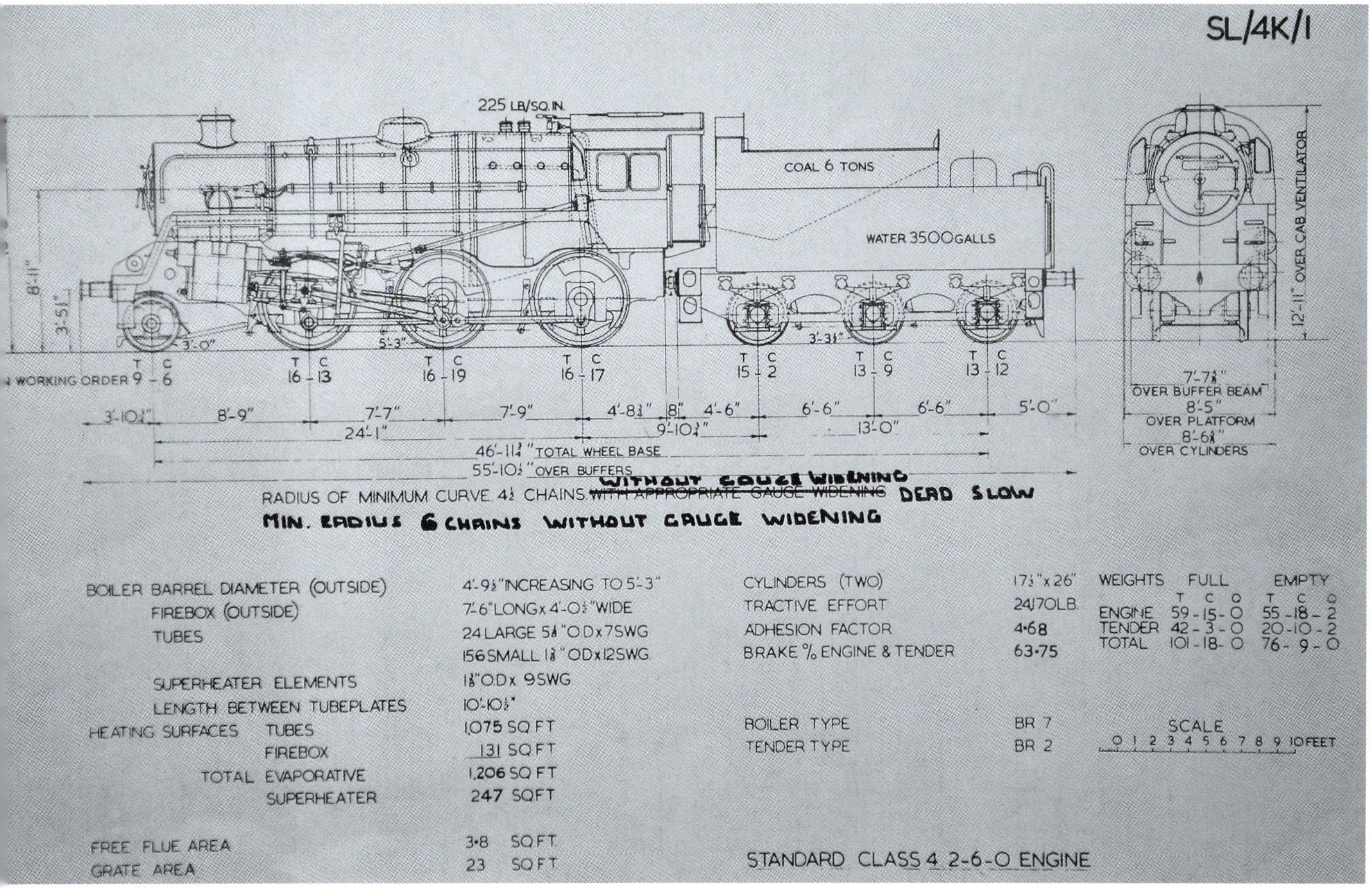

# A Long Goodbye • 219

[Engineering diagram of Standard Class 4 2-6-0 engine with annotations SL/4K/1, showing 225 LB/SQ.IN., COAL 6 TONS, WATER 3500 GALLS, and various dimensions. Handwritten annotations read "WITHOUT GAUGE WIDENING", "WITH APPROPRIATE GAUGE WIDENING DEAD SLOW", and "MIN. RADIUS 6 CHAINS WITHOUT GAUGE WIDENING".]

| | | | | | |
|---|---|---|---|---|---|
| BOILER BARREL DIAMETER (OUTSIDE) | 4'-9½" INCREASING TO 5'-3" | CYLINDERS (TWO) | 17½" x 26" | WEIGHTS | FULL EMPTY |
| FIREBOX (OUTSIDE) | 7'-6" LONG x 4'-0⅜" WIDE | TRACTIVE EFFORT | 24,170 LB. | | T C O T C O |
| TUBES | 24 LARGE 5¼" O.D x 7 SWG | ADHESION FACTOR | 4·68 | ENGINE | 59-15-0  55-18-2 |
| | 156 SMALL 1⅞" OD x 12 SWG | BRAKE % ENGINE & TENDER | 63·75 | TENDER | 42-3-0  20-10-2 |
| SUPERHEATER ELEMENTS | 1⅛" O.D x 9 SWG | | | TOTAL | 101-18-0  76-9-0 |
| LENGTH BETWEEN TUBEPLATES | 10'-10½" | | | | |
| HEATING SURFACES  TUBES | 1,075 SQ FT | BOILER TYPE | BR 7 | | |
| FIREBOX | 131 SQ FT | TENDER TYPE | BR 2 | | |
| TOTAL EVAPORATIVE | 1,206 SQ FT | | | | |
| SUPERHEATER | 247 SQ FT | | | | |
| FREE FLUE AREA | 3·8 SQ FT | | | | |
| GRATE AREA | 23 SQ FT | STANDARD CLASS 4 2-6-0 ENGINE | | | |

returning with the 7.05pm service the same day. Its success led to other Eastleigh based locomotives running regularly over the line on this service, beginning later in March. Very soon, their use was expanded to cover other passenger duties including the Pines Express, though in this case it was thought necessary for such a heavy train to need the support of a 2P, as witnessed by Ivo Peters' many photographs. The Class 4 may not have needed assistance, but it was rumoured at the time that these engines were descended from Ivatt's 4MT Class 2-6-0s which had performed so poorly when attached to the S&D. So it may have been the

**The Doncaster** designed Standard Class 4 2-6-0 became a regular performer over the S&D from March 1955. Their design is recorded as owing much to Ivatt's 4MT 2-6-0s that first appeared in 1947, which were later described by Ivo Peters as 'originally designed with a double blastpipe and chimney and performed abysmally'. So, perhaps, the new 2-6-0s arrived under a cloud of low expectation. However, Peters then recorded that these standard engines were 'quickly accepted by S&D enginemen as good, reliable locomotives'. (Top) – A diagram of the type which Peter Pike, then a fitter at Templecombe, kept amongst his papers, with later annotations. (Below) Engine No. 76066 which became a regular performer on the S&D here photographed ambling through Bournemouth Central Station in the early 1960s. (Author)

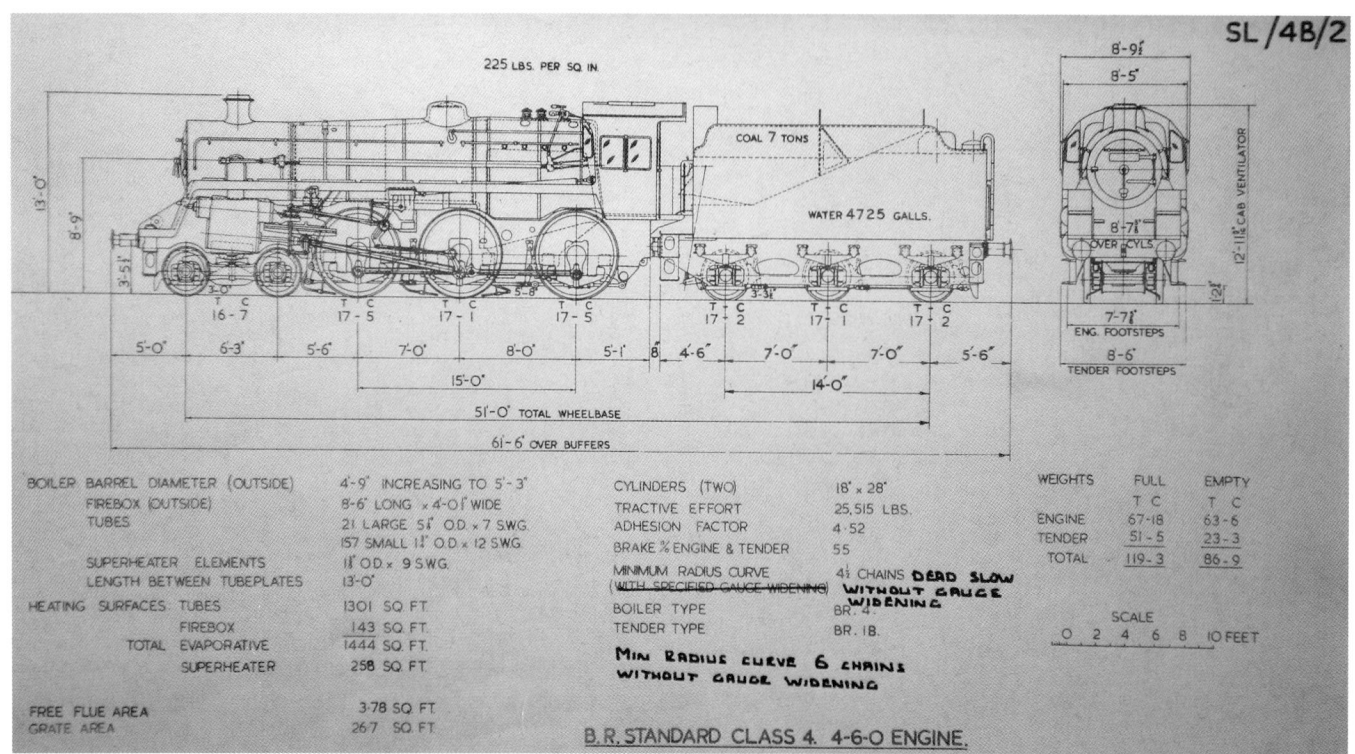

(Top) The Standard Class 4 4-6-0 as designed at Brighton and (below) No. 75073 on duty at Templecombe in the early 1960s, by which time the condition and cleanliness of many steam locomotives had deteriorated badly as cuts and an inability to recruit staff began to bite. (Author)

case, as Ivo Peters surmised in his books, that some prior knowledge and the application of common sense dictated a safety first response until the new class had proved itself over such a demanding line. After that, it is recorded that at least seven of the class regularly served on the S&D.

The 5s and 4s were soon joined by a third group of standard engines. By 1955, production of the Class 4 4-6-0 mixed traffic locomotives at Swindon was approaching sixty, with twenty more to be added by May 1957. This fairly large number meant that these lightweight versions of the Class 5s, which were designed by the draughtsmen at Brighton under Ron Jarvis, found fairly wide use across the network. Their low axle loading also allowed them to operate where the 5s could not go and for that matter where engines with a similar power rating, most notably the WR's Manor Class, which were wider over their cylinders, were restricted. Despite being a product of Brighton's drawing office, the design owed much to the 2-6-4Ts devised at Derby by Tom Coleman and his team when Charles Fairburn was CME, some of which were built at Brighton in 1950/51 for use on the Southern. These, in turn, had evolved from the effective and reliable two-cylinder 2-6-4Ts that appeared in 1935 under Stanier. The key inherited feature seems to have been the boiler, but in this case, the barrel was lengthened by nine inches on the front parallel ring, with the firebox being virtually the same size. In addition, the cylinders, much of the motion, the boiler mountings, the total heating surface, wheels and bogies were also very similar. However, the Class 4's boiler pressure was set at 225 psi by comparison to the 2-6-4's 200 psi and it produced a tractive effort of 25,515lbs, against 24,670lbs. In service, the Class 4s proved to be a successful design and appear to have gained a good reputation amongst footplate crew, with one later writing that 'it was one of the more effective BR standard types'. Was this a back-handed compliment, suggesting some of the others weren't up to much? We shall never know, but these new engines do seem to have been well-received and in 1955 the S&D

*Below and overleaf*: **Peter Pike** began his career as an apprentice, then fitter at Bath in the mid-1950s before becoming a fitter at Templecombe later in the decade. He has left a day to day account of the work in which he was involved. These logbooks may well be one of the few records of this type to survive. Four pages from the 1958 volume are included here to demonstrate the variety of work undertaken, the locomotives he dealt with that year and the various shedcodes of engines, from across the country and different regions, that passed over the line. For such a small railway, the mixture of shedcodes and the variety of locomotives this demonstrated is quite extraordinary. (P. Pike/Author)

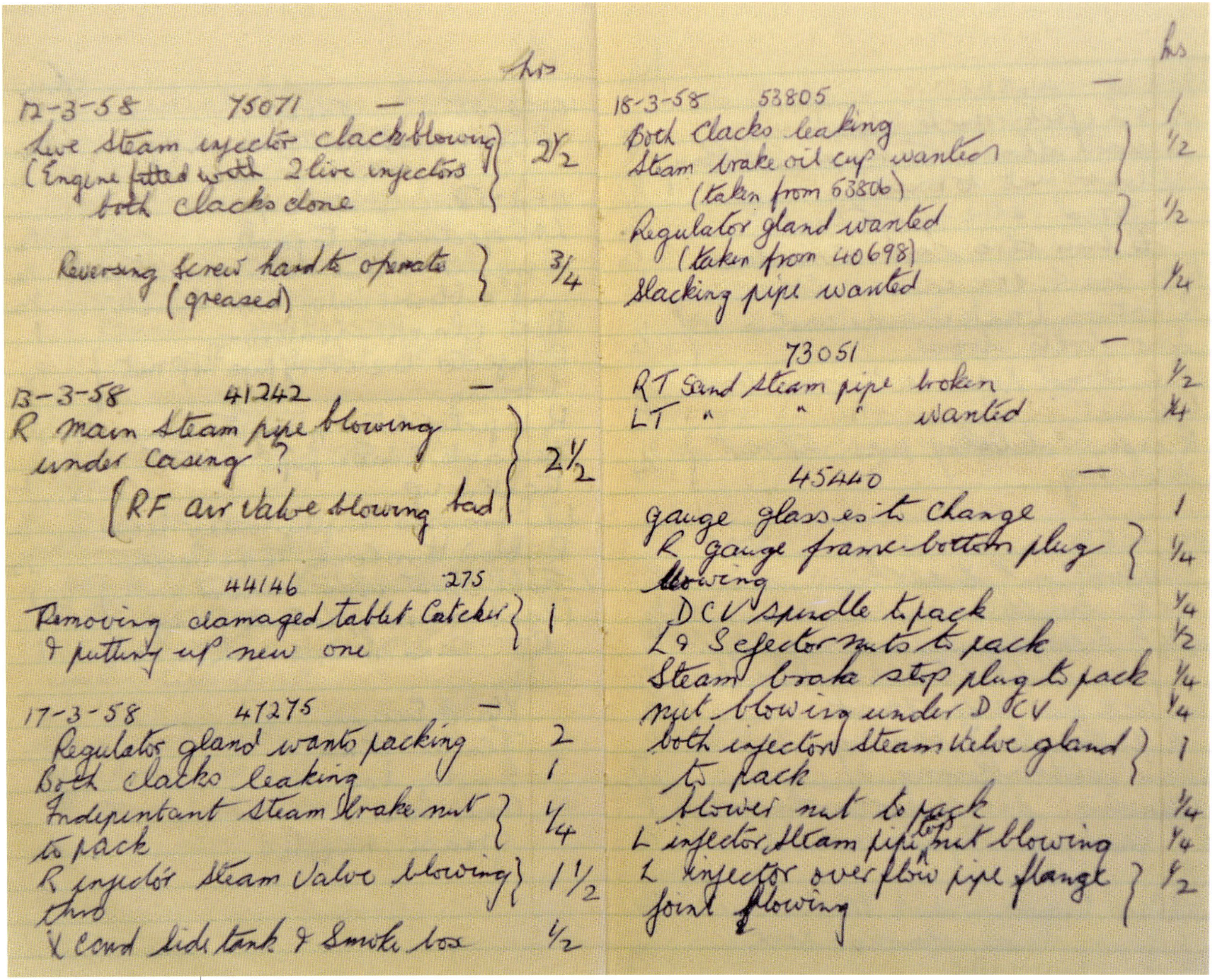

| | | hrs | | | | hrs |
|---|---|---|---|---|---|---|
| 12-3-58 | 75071 | — | 18-3-58 | 53805 | — | |
| Live steam injector clack blowing (Engine fitted with 2 live injectors) both clacks done | | 2½ | Both Clacks leaking | | | ½ |
| | | | Steam brake oil cup wanted (taken from 53806) | | | ½ |
| Reversing screw hard to operate (greased) | | ¾ | Regulator gland wanted (taken from 40698) | | | ½ |
| | | | Slacking pipe wanted | | | ¼ |
| | | | | 73051 | — | |
| 13-3-58 | 41242 | — | RT Sand steam pipe broken | | | ½ |
| R main steam pipe blowing under casing | | 2½ | LT " " " wanted | | | ¼ |
| (RF air valve blowing bad) | | | | 45440 | — | |
| | | | gauge glasses to change | | | 1 |
| | 44146 | 275 | R gauge frame bottom plug blowing | | | ¼ |
| Removing damaged tablet catcher & putting up new one | | 1 | DCV spindle to pack | | | ¼ |
| | | | L & S ejector nuts to pack | | | ½ |
| 17-3-58 | 47275 | — | Steam brake stop plug to pack | | | ¼ |
| Regulator gland wants packing | | 2 | nut blowing under DCV | | | ¼ |
| Both clacks leaking | | 1 | both injectors steam valve gland to pack | | | 1 |
| Independant steam brake nut to pack | | ¼ | blower nut to pack | | | ¼ |
| R injector steam valve blowing thro' | | 1½ | L injector steam pipe nut blowing | | | ¼ |
| | | | L injector overflow pipe flange joint blowing | | | ½ |
| X coal side tank & smoke box | | ½ | | | | |

were allocated three ex-Exmouth locomotives – Nos. 75071, 75072 and 75073. Others were occasionally spotted on the line and in 1960, engine No. 75027 was allocated to Templecombe.

1958 saw the organisational changes mentioned earlier take effect and the WR gain control of the line as far south as Henstridge. Perhaps more importantly, the bulk of the locomotive fleet became the responsibility of the Western Region as did the main sheds at Bath and Templecombe and the sub-sheds at Radstock and Highbridge. However, four became three on 11 May 1959 when Highbridge closed, except for

**On 15 January** 1958, the S&D suffered one last major accident, on this occasion involving a loose-coupled freight train running from Highbridge to Templecombe with thirty-two wagons containing a mixed load of coal and tinned milk, plus a guard's van. In heavy rain and a rapidly descending mist, Driver Ron Spiller and Fireman Peter Guy, on Armstrong Whitworth 0-6-0 No.44557, were struggling up the bank from West Pennard towards Pyle. With wheels slipping, the engine struggled towards Pyle Station when a wagon coupling parted leaving only one truck attached to the engine. The rest began to run back down the incline rapidly gathering speed. The signalman at West Pennard had the time and presence of mind to switch the train into sidings adjacent to the station, rather than let them run on down the line to Glastonbury. The speeding train, slowed slightly by the action of the guard, Ted Scovell, in his van, entered the sidings, derailed and wrecked itself. The guard, in a deeply shocked state, emerged virtually uninjured from the wreck, but it was said that he never recovered from the shock. The photograph above was taken the following day as the clear up begins. (P. Pike/Author)

overnight stabling when required. With the new management came a number of new initiatives. Here the WR sought to improve the line's motive power or, alternatively, find work for under-employed engines from across its region as diesels began to make their presence felt. This involved trials during December 1958 with Class 2251 0-6-0 tender engine No. 2215 and two types of Pannier tank locomotives, a Class 5700 0-6-0, No. 3604, and a Class 9400, No. 8451. Each trial seems to been successful and engines of these types soon became fixtures, in fairly small numbers, on the line. Less successful was the attempt to use a Riddles wartime built Austerity 2-8-0 engine, No. 90125, as a way of meeting a need created by the withdrawal of the S&D's 7Fs. On a single run with a mixed goods train from Bath to Evercreech during the morning of 22 January 1959, its brakes were found to be inadequate for the task

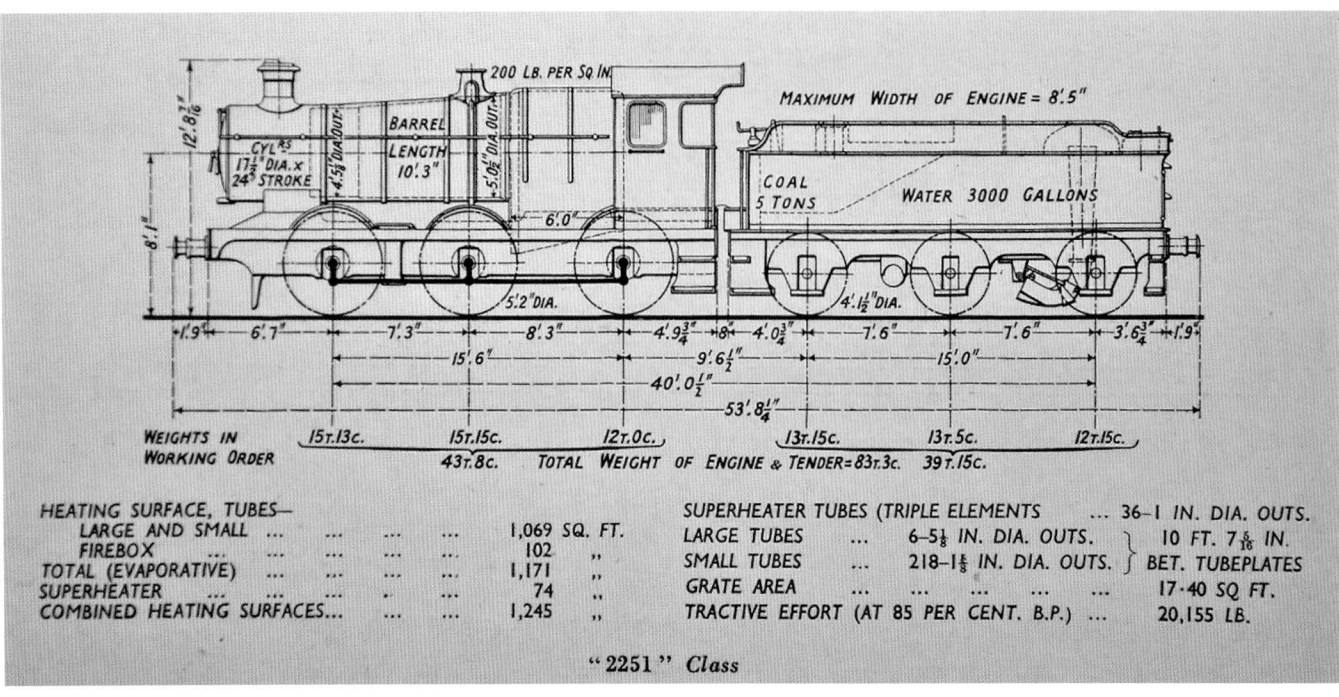

**The Class** 2251 0-6-0 tender mixed traffic engines first appeared in 1930 and were built to serve the lines where axle loading restrictions prohibited the passage of large locomotives. Under Charles Collett, Swindon designed the two-cylinder 2251s and between 1930 and 1948 the company built 120 at Swindon. The frames and motion were almost identical to the Class 5700 Pannier tank engines. The tapered boiler was a modified GWR standard No. 2 and in service they soon gained a reputation for steaming well, versatility on both goods and passenger services and steady running, coupled to wide route availability. As such, their allocation to the S&D must have seemed a sensible one and in the few years left to the line at least five served there – Nos. 2218, 3200, 3201, 3205 and 3210, with the fourth of these surviving into preservation. (Top) The overall dimensions of the class and (below) engine No. 3201 photographed at Glastonbury on 3 April 1965. (Author)

| | | | |
|---|---|---|---|
| HEATING SURFACE, TUBES— LARGE AND SMALL ... | 1,012·8 SQ. FT. | TUBES ... { 2–5⅛ IN. DIA. OUTS. / 219–1⅝ IN. DIA. OUTS. } | 10 FT. 6 13/16 IN. BET. TUBEPLATES |
| FIREBOX ... | 102·5 ,, | GRATE AREA ... | 15·3 SQ. FT. |
| TOTAL ... | 1,115·3 ,, | TRACTIVE EFFORT (AT 85 PER CENT. B.P.) ... | 22,515 LB. |
| NO SUPERHEATER | | | |

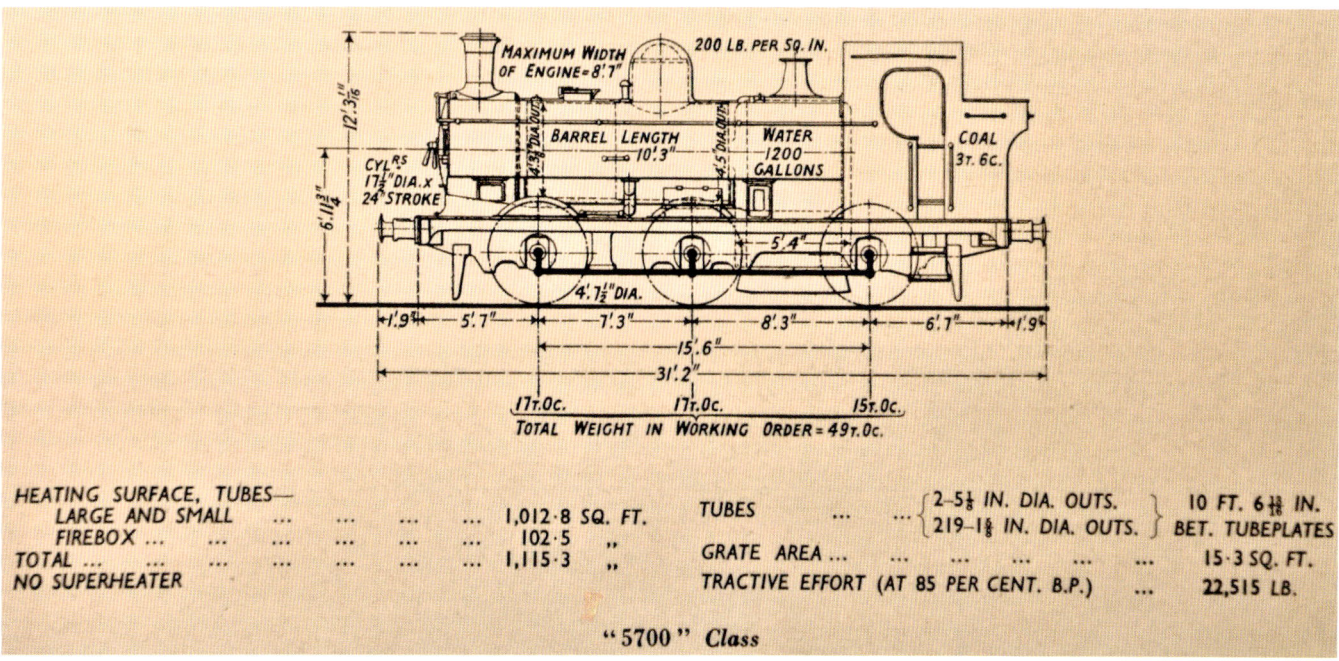

"5700" *Class*

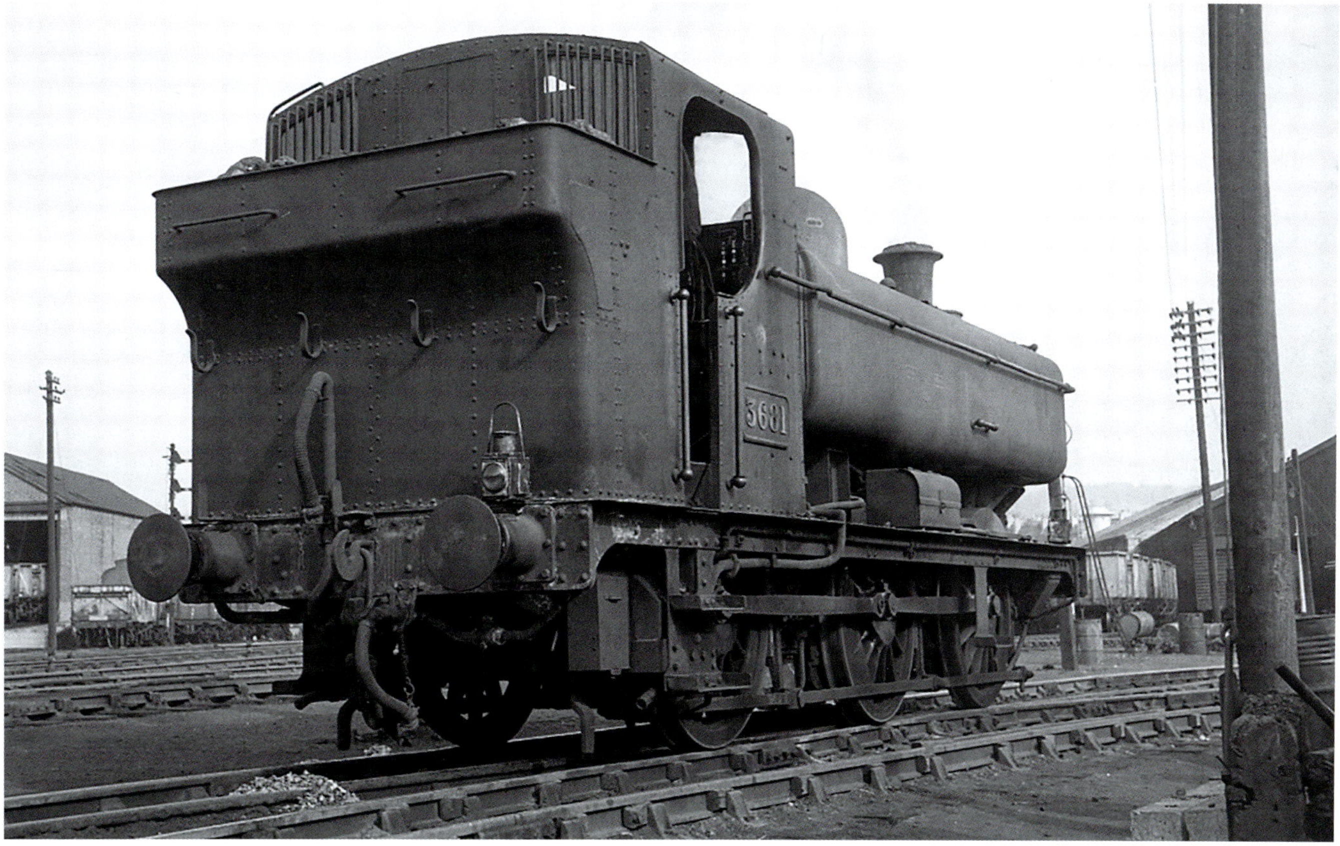

**The GWR** had a long history of designing and building pannier 0-6-0 tank engines. The programme began in 1864 with the introduction of eight Class 302 side tank engines, seven of which were rebuilt as panniers later. The same development plan then swept up the 1874-introduced Class 850 saddle tank programme, which produced 170 locomotives by 1895 with many of these being rebuilt as panniers later. Development continued, most notably with the conversion of the 2721 Class that first appeared in 1897. (Top)The 5700 Class, as shown in the diagram above, were a later incarnation of the type and between 1929 and 1950 863 of these successful freight tank engines were built, under Charles Collett's leadership. Some of them worked on the S&D post-1958, including Nos.3681, 3742, 4631 and 4691. (Below) The first of these engines is captured here on 3 April 1965 at Bath. (Author)

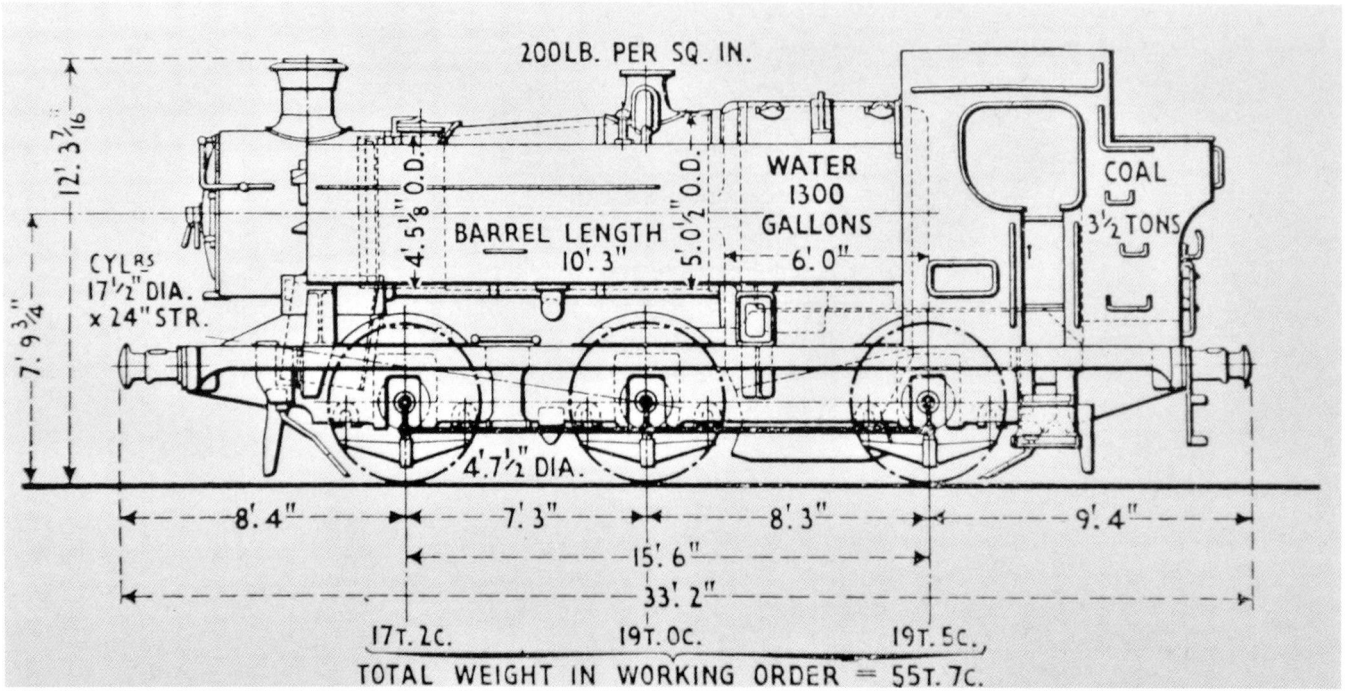

**(Top) Another** version of the pannier tank to appear on the S&D were the 9400s which entered service in 1947. By 1956, 210 of these heavy shunting locomotives had been built. In essence, as the diagram confirms, they owed much to the 5700 Class, but modified, under Frederick Hawksworth when CME, to include, amongst other things, a No. 10 tapered boiler and firebox, a drumhead smokebox and a wider cab. Though not deemed as successful as the 5700s, these engines still performed effectively with several, according to Peter Pike's records, working on the S&D, including No. 8486 (lower picture), a Robert Stephenson and Hawthorn 1952 built engine. It is captured here on shed in Bath in the mid-1960s, with hand painted smokebox number plate. Presumably, the original plate had been removed to stop it being 'souvenired' by a collector as the line and steam approached their end. (Author)

and the train 'ran away', according to contemporary reports. With safety so easily compromised, it returned to Bristol and this encouraged the WR to seek the return of the reliable LMR 8Fs. At about the same time, a GWR 56XX 0-6-2T appeared, but, like the Austerity engine, was found unable to control its load. In this case it was a morning coal train from Midsomer Norton to Bath, and like the Austerity 2-8-0 it too quickly disappeared.

For those working on the railways, the 1960s seemed to promise very little. The growth of road vehicles had drawn many of the railway's traditional customers away and this loss grew rapidly as the motorway network gradually expanded. BR tried to respond but found cutbacks and closures a dominating force that could not be ignored. Dieselisation and electrification helped stem the haemorrhage of customers, but the freedom offered by cars, buses and lorries had an ever-growing attraction. In March 1961, the debate over the level of competition BR faced came to a head when Dr Richard Beeching, physicist and engineer, was appointed Chairman of BR's Board. His brief was a very simple one; rationalise and modernise, cut away the deadwood and, in theory, improve BR's ability to compete. But his work would take time to reach fruition and, whilst that happened, each region struggled on, making best use of the tools at their disposal.

The GWR and then the Western Region are often portrayed as enemies of the S&D, eager to see its closure and actively planning to do so. If this is true or not is a matter of opinion and conjecture, but at least, in the early 1960s, they boosted the line's fleet of locomotives with two of the best standard class engines to be built. The first 9F 2-10-0, No. 92204, arrived in March 1960 and then a Class 4 2-6-4 tank locomotive, No. 80081, appeared in late 1963. Two eleventh hour additions and perhaps, to the optimistic, a sign that the line might survive the difficult times that lay ahead.

The 9F was, in many people's eyes, the ideal locomotive for the S&D and was greeted with great enthusiasm by footplate crew. Their reaction was probably best summed up by Peter Smith, in his book *Mendips Engineman,* when recalling seeing engine No. 92205 at Bournemouth. He and driver

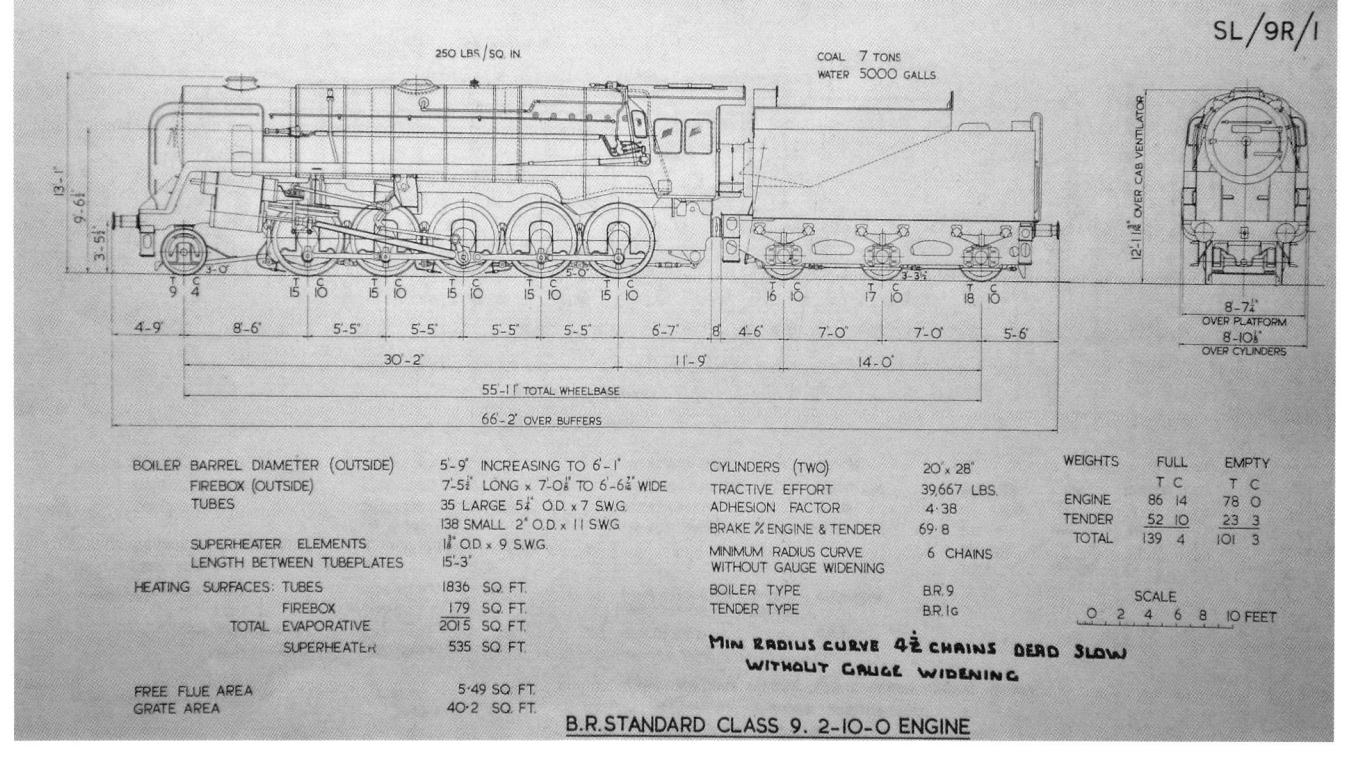

**Two hundred** and fifty-one 9F 2-10-0s were built between 1954 and 1960, mostly at Crewe but a number of others at Swindon as well. However, design work was undertaken by Ron Jarvis and his team at Brighton. There were some variations in the types produced; some had double blastpipes, ten were fitted with Crosti boilers and heating equipment, two with mechanical stokers and one, No. 92250, with a Giesl oblong injector. There were also five different types of tender designed to suit varying regional requirements. This diagram captures one variant. (Author)

Donald Beale were due to take her on the Pines Express to Bath and he simply recorded that it was 'for Donald and myself a case of love at first sight'. High praise indeed from two S&D stalwarts. After a successful run, during which 'Donald put a few questions to her at various locations and she answered them all with obvious enjoyment', he concluded that 'we were vastly impressed with this [its strength, steaming qualities, good coal consumption, smooth riding and much more], as with every aspect of this incredible locomotive'.

During 1959/60, the cost of running the S&D was continuing to cause concern to the WR's senior managers and savings were always being sought to reverse this trend. There was also the issue of staffing problems to consider. With fewer men coming forward to serve on the railways, it was becoming increasingly difficult to keep the service going, particularly when it came to labour intensive steam locomotives. As a result, reducing the amount of double-heading, without cutting the loads pulled, was one obvious way of overcoming these problems. The 9Fs could provide one answer short of closure and had proved their capacity to do so across BR's network in six years of operation. But first a trial was necessary, hence St Philip's Marsh based 92204's appearance at Bath on 29 March. This engine was one of a batch of eighteen built at Swindon in 59/60 so was almost fresh from the production line and would have been in excellent condition. So it wasn't a surprise that it coped well with an eleven coach train weighing 350 tons on a return run between Bath and Bournemouth unassisted throughout, despite heavy rain, high winds and snow flurries. Armed with this information, the WR decided to allocate four locomotives, Nos. 92203 to 92206, to the S&D for the summer season that year.

The same routine was followed in 1961, with engines 92000, 92001, 92006 and 92212, plus three occasional visitors, 92059, 92078 and 92152. 1962 saw the process repeated but this time the allocation was accompanied by a sense of a finality which seemed to clarify many people's concerns over the future. For the summer season, engines 92001, 92010, 92233 and 92245 appeared, but there would be one important addition – 92220 *Evening Star*. Its arrival was timed to coincide with the withdrawal of through services to and from the north, which were to be routed via the ex-GWR line through Birmingham and Oxford to the south coast. This included the Pines Express, its last run on 8 September being taken by 92220, appropriately the last steam locomotive built for BR and destined to be preserved as part of the national collection. However, this engine, with No. 92224, returned to the line for three months in 1963 when there was a shortfall in motive power with other engines under repair. Two Black Fives were requested, but the 9Fs arrived instead. By this time, passenger traffic over the S&D had reduced significantly and these two massive engines found employment pulling three or four carriage trains. But this was only part of a general rundown that had begun in 1958 and grew in pace into the new decade. Gradually freight traffic was also routed away from the S&D and other passenger services

**From any** angle, but particularly this one, the 9Fs were impressive, with looks that translated into sheer power. It is to BR's everlasting shame that these impressive engines, with useful lives that could have stretched into the 1990s, were scrapped after only a few years. Better, perhaps, if they and the other standard engines had never been built and the political will existed to allow the science of diesels to be exploited much sooner as in other advanced nations. Nevertheless, while the 9Fs lasted they were a fitting memorial to steam locomotion. (Author)

reduced in scale. The outstanding 9Fs and their demeaning role in 1963, followed by a brief walk on part in June 1964 with No. 92214 and then their final departure with No. 92238 in June 1965, pulling an enthusiasts' special, seemed to sum up the line's sad predicament and their going became its swansong.

Before the final curtain fell on the S&D, BR's Class 4 2-6-4 tank engines made their appearance. The first arrived on 4 November 1963 and was often found, as the year came to an end, working three or four carriage stopping services between Bath and Bournemouth, a service better served by the 2-6-4Ts than the 9Fs. It also double-headed

**In 1960,** engine No. 92203, built at Swindon in 1959, was one of four locomotives assigned to the S&D for the summer months. This 9F, which is photographed at Liverpool, did not share the line's fate and survived into preservation, being bought by the artist David Shepherd and then named *Black Prince*. (Author)

**No. 92220** *Evening Star*, photographed at Swindon on 21 March 1964. It entered service exactly four years before this picture was taken and was withdrawn in March 1965 for preservation. It was allocated to Bath Green Park twice, from August to October 1962, when she pulled the final Pines Express, and again at the end of August 1963 until November. During this second tour of duty, she would have little to do because by this stage, the Western Region had re-routed all traffic from the north away from the S&D so leaving her to face the ignominy of pulling three or four carriage trains between Bath and Bournemouth, such was the state of steam engines at this time and the rapidly advancing modernisation of BR's diesel and electric fleet. (Author)

**The Brighton** designed Class 4 2-6-4T first appeared in 1951 and by 1956 155 had been built at Derby, Brighton and Doncaster. The design had clear antecedents back to Stanier's 1935 two cylinder engines of the same class and their Fairburn derivatives of 1945. But Jarvis and his team had to refine the design to meet the universal L1 loading gauge. This was achieved by reducing the cylinders from 19⅝in x 28in to 18in x 28in, and by modifying their overall structure. In addition, standard BR components were fitted where possible and the boiler's pressure was raised from 200 to 225 psi. In service, these engines proved to be successful, achieved wide route availability over the network and, apparently, were popular with footplate crew. They were also one of my favourites, as I rode behind them on many occasions, especially on summer holiday trips to Broadstairs. (Author)

**Although this** engine is not recorded as having served on the S&D, this photograph does at least capture the Class 4 2-6-4T's balanced and elegant design. It was probably a class that could have served the S&D well if allocated earlier in their lives. As it was, they proved a very useful addition to the slowly diminishing fleet during the last three years of the line's life. (Author)

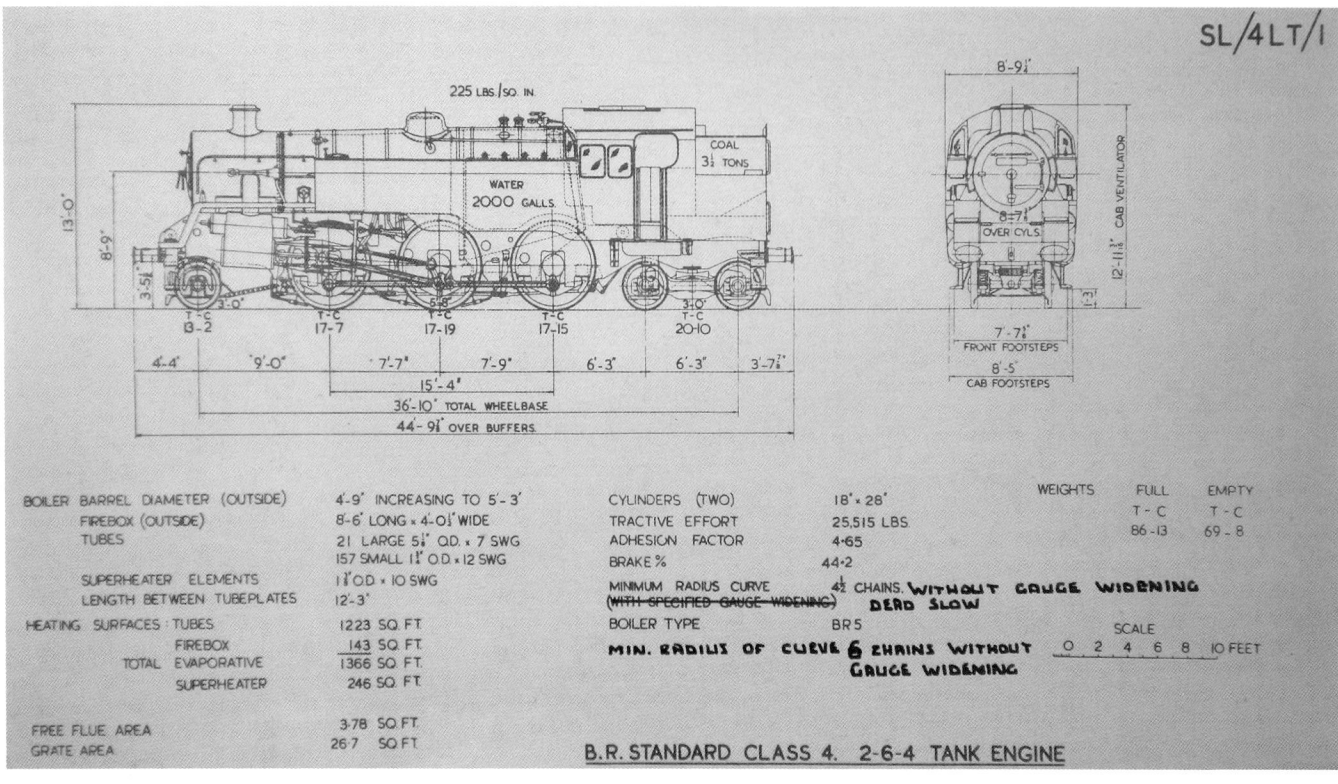

any remaining passenger services where the load might exceed seven or eight carriages. In May 1964, engine No. 80081 was joined by Nos. 80043, 80059, 80067 and 80146, and by 1966 this had increased by two more, Nos. 80039 and 80138, to make them a very common sight as the S&D moved, inexorably, towards closure.

The last two years of the S&D's life was a death in slow motion, with any attempt to reverse the downward trend met with the unflinching face of bureaucracy and the bleating cry of 'it's losing too much money'. It is very easy to believe the conclusion that some have reached that this situation was manufactured by BR's managers, most specifically those at Swindon, to meet a much broader political aim. And the way the line was gradually stripped of its more profitable services lends weight to this argument. However, many areas of Britain's railway network were in a similar position at this time and were barely able to attract sufficient business to survive without subsidies. With the growing dominance of road vehicles, few in the 1960s would have gambled on a railway revival.

So streamlining and modernisation became the overriding business plan and reduction of services was seen as crucial to achieving this goal. When the Beeching Report was finally published on 27 March 1963, and approved in the months that followed, it underpinned all this, and the S&D and many other parts of BR were finally marked down for closure. Salvation was expected when the Labour Party took office in October 1964, but their commitment to the railways proved to be an illusory one, and the process of dismemberment continued. During the autumn of 1964, overnight freight trains and mail services were withdrawn, and by day many goods trains were re-routed across other lines. In these declining years, the lack of funds and interest at senior level was reflected in the deteriorating condition of the line's infrastructure and its locomotives. Dirt and unchecked decay became the order of the day and a once proud company and its dedicated staff became hostages to a rapidly changing world. The chop finally fell on 10 September 1965, with closure set for 3 January the following year. But replacement road services could not be put in place quickly enough and so there was a two month stay of execution until the beginning of March, when, finally, all was lost.

**Standing cold** in front of the old Midland Engine Shed at Bath in 1965 or 66. An unidentified 8F, a Standard Class 3 2-6-2T No.82004 and Class 4 No. 80059 seem to have been discarded such is their lifeless condition and state of cleanliness. (Author)

In the last years of the S&D, a slow deterioration was inevitable. Budgets were cut and staffing levels reduced. The decline was particularly apparent in the condition of the engines, where minimal maintenance was practised and once shining engines became increasingly grimy. Across BR, when an engine broke down it was quite likely to be cannibalised for spares to keep others going or be scrapped. With the rapid withdrawal of steam in the 1960s, there were many other engines to take their place, so it was an easy policy to implement. This group of photographs captures these final years on the S&D and the efforts made by its dedicated staff to keep the line going. (Above left) – Two West Country Pacifics on shed at Templecombe, or so the notes attached to the negative suggest. No 34093 *Saunton* was based at Eastleigh for the last years of her life but was a regular on the S&D for many years. She was withdrawn in July 1967. The second engine is unidentified but has been stripped of its plates and motion suggesting that it might have been withdrawn already and is awaiting its final journey to the scrapyard. (Below) – Stanier's 8Fs were present on the line until the end and pulled the occasional special train in this case engine No. 48309, specially cleaned for this duty, is in charge. (Author)

(Left) An unidentified Class 4 2-6-4T on the outskirts of Bath with a stopping train made up of three Southern Region green carriages on 2 April 1965. (Below) 0-6-0T No. 47276 working at Writhlington on 3 June 1966. This engine was not on allocation to the S&D on a permanent basis but seems to have been a casual visitor. (Author)

The last part of the line's life was fraught with uncertainty and anxiety, but during these difficult years, footplate crew and shed staff struggled on as best they could. In so doing they had to handle a very varied fleet of locomotives, not all on permanent allocation to the line. Many of these would have been in an increasingly jaded state as BR speedily ran down the S&D and steam locomotives in general. The variety was reflected most graphically in the visiting engines called in to haul trains over the line before it closed, many of them enthusiast specials. The list reads like a history of locomotives over the last thirty years of steam development. Amongst others there were two Merchant Navy Pacifics, a Britannia Class Pacific, No. 70034 *Thomas Hardy*, a GWR Castle and then a Manor, two Royal Scots, a 4-6-0 Patriot, two B1 4-6-0s and the Horwich Moguls. Even some diesels appeared, including the Sulzer Type 4s and Class 35 Hymeks, to point the way to an alternative future denied to the S&D by closure.

By 1966 there were only two older engines left from the line's pre-war days. All the 7Fs had gone, the last in October 1964, although two survived into preservation. The last 0-6-0 tender engine built by Armstrong Whitworth, No. 44560, went in August 1965, while two 1929 W.G. Bagnalls built 0-6-0Ts, Nos. 47314 and 47312, ran on until November 1966 and June 1967 respectively. To this could be added the last 4-4-0 in service, No. 40634, which went in May 1962, the 1902 Neilson Reid built 0-6-0, No. 43216, withdrawn in August 1962, and the last Radstock shunter, Sentinel 0-4-0T No. 47190, which had struggled on until March 1961. Even the newer arrivals, amongst them Bulleid's Pacifics, Stanier's Black Fives and 8Fs and the BR standards, were soon scrapped as the cull of steam locomotives across

**(Opposite) 4-4-0** No. 40537 at Templecombe in July 1960; once again this engine appears to have been on temporary loan to the S&D.
**(Above)** The same could not be said of Armstrong Whitworth 0-6-0 No. 44560 which was built for the S&D in 1922 and remained on the line until withdrawn in August 1965. This photograph was taken at Templecombe shed on 30 March 1964. (Author)

(Above) 8F No. 48760 receives some attention at Bath in April 1965 flanked by an unidentified ex-GWR Pannier tank and BR Class 4 2-6-4T. (Below) Standard Class 4 2-6-0 No. 76062 comes to rest at Green Park on 13 October 1964 to a scene of low activity, perhaps summing up the state of the S&D at this stage. (Author)

**The sheds** at Bath were always a fascinating and atmospheric place to visit especially in the last two years of the line's life, as witnessed in this photograph taken in October 1964 with 8F No. 48706 the only identifiable engine. (Author)

Britain gathered pace. All were withdrawn from main line service by 1968, when many still had years of productive life left in them. Such was the arbitrary, poorly thought out nature of this dire programme.

There is now very little left of the Somerset and Dorset to remind the casual observer that it had once existed. When writing this book, I took the opportunity to visit what now remains of the line. BR and its contractors did a very good job in trying to hide any sign of its existence. Like an archaeologist, I ended up rooting around in the undergrowth and looking for signs of disturbed earth where the track or buildings once lay. Many structures were literally blown up or demolished, while others were taken over by other businesses or became homes. But the culture formed by the nature of the line and its many employees over a hundred years of life was another matter. This was destroyed in an instant, but while a few men and women the railway employed survive, so a living memorial exists. But this link will soon be gone and then we will be left with books, photographs and some cine film to remind us that the S&D once existed. Perhaps the only living memorial we have is Bath Green Park Station itself, which found a new life as a car park, with shops and cafes occupying its rooms. It is still a joy to walk through its portals on what remains of the wooden platform especially on market days when the concourse is full and very busy. Luckily, sufficient is left for me to recall my life as a teenager when after school I would amble through the station, stop to watch the trains, then stroll on a little further to catch a 339 bus home, regretting that I never had the funds to ride a train over the Mendips.

**Quite a** busy scene at Green Park in 1964 as Standard Class 5 No.73004 makes ready to leave. The attention the engine is receiving suggests that this might have been a 'special' for enthusiasts. (Author)

**I wanted** to end this book with a single photograph that to me summed up the life of the S&D. I could find none better than this evocative shot of a 7F, No.53807, hissing with steam as it pulls away. It is a quintessential British scene common to those who lived in the 1950s and 1960s before science and a desire for modernisation moved us on, as it always does. (D. Neal)

# REFERENCE SOURCES

**Archives/Collections Consulted**
T.F. Coleman/ M. Lemon Collection.
R. Hillier Collection.
Institution of Mechanical Engineers.
National Archives (Discovery).
National Railway Museum.
D. Neal Collection.
P. Pike Collection.
R.A. Thom Collection.

**Books and Other Publications Consulted**
ARLETT, M., *The Somerset and Dorset at Midford,* Millstream Books (1988).
ATTHILL, R., *The Somerset and Dorset Railway,* David and Charles (1967).
BARRIE, D.S. & CLINKER, C.R., *The Somerset and Dorset Railway,* Oakwood Press (1978).
BRADLEY, D. & MILTON, D., *Somerset and Dorset Locomotive History,* David and Charles (1973).
*The Engineer* Magazine – various issues.
HAMILTON-ELLIS, C., *The Midland Railway,* Ian Allan (1953).
HAMMOND. A., *Reminiscences of the Somerset and Dorset,* Millstream Books (1997).
HARESNAPE, B., *Fowler Locomotives,* Ian Allan (1972).
HARESNAPE, B. *Stanier Locomotives,* Ian Allan (1970).
HAWKINS, M., *The Somerset and Dorset Then and Now,* Guild Publishing (1986).
*ILocoE Journal* – various issues.
*IMechE Journal* – various issues.
JUDGE, C.W. & POTTS, C.R., *An Historical Survey of the Somerset and Dorset Railway,* OPC (1979).
*Meccano Magazine* – various issues.
PETERS, I., *The Somerset and Dorset in the Fifties – Volume One,* OPC (1980).
PETERS, I., *The Somerset and Dorset in the Fifties – Volume Two,* OPC (1981).
PETERS, I., *The Somerset and Dorset in the Sixties – Volume Three,* OPC (1982).
PETERS, I., *The Somerset and Dorset in the Sixties – Volume Four,* OPC (1982).
*The Railway Gazette* – various issues.
*Railway Magazine* – various issues.
SMITH, P., *Footplate Over the Mendips,* OPC (1978).
SMITH, P., *Mendips Engineman,* OPC (1972).

# INDEX

George England 2-4-0T locomotive No. 8 was built in 1861, but was, a few months later, rebuilt as a saddle tank, as pictured here. It was eventually withdrawn from service in 1928 having been renumbered twice, as 28 then 28A, but underwent three more transformations in the process. It became a 2-4-0 for a seven year period beginning in 1876, then a further 21 years as a saddle tank engine again, before reverting to a tank loco in 1904 in which state it reached the end of its life. (R Hillier)

Adams, William – 49, 78, 192.
Allport, James – 24.
Anderson, James – 75, 124, 125, 137, 144, 149, 150, 156, 157, 179.
Andrews, Robert – 22, 59.
Archbutt, Reginald – 145, 151, 157, 170, 175, 178.
Argentine Railway – 134.
Armstrong Whitworth – 151, 152, 153, 182, 223.
Aspinall, John – 130, 131.
Association of Locomotive Engineers (ARLE) – 150, 151, 158.
Atthill, Robin – 9, 16, 35, 44, 46, 103.
Avonside Engineering Co – 88-96.

Baedeker Raids – 44.
Bagnall, W G – 170.
Baldwin Works – 136.
Barlow, W H – 21.
Bath (incl Green Park) – 8-10, 19-21, 24, 26, 29, 36, 37, 40, 44-46, 69, 114, 115, 118, 127, 128, 146, 162, 178, 182, 191, 195, 198, 203, 208, 210, 213, 214, 222, 226, 228, 229, 233, 236-238.
Bath accident (1929) - 167-169.
Beale, Donald – 228.
Beattie, Joseph – 49, 53, 55
Beattie, William – 49, 53,
Beeching, Richard – 46, 227, 231.
Belgium Railway – 138.
Belpaire, Alfred – 96.

Belpaire firebox – 107, 108, 127-129, 153, 170.
Beyer Peacock – 125.
Billington, John – 156.
Binegar – 20,
Birmingham and Gloucester Junction Railway – 74.
Bishop's Castle Railway – 58.
Blandford – 15, 19, 37, 185.
Board of Trade – 66.
Bond, Roland – 203, 204.
Bournemouth – 114, 115, 213, 214, 219, 228.
Brankstone – 37, 213.
Bridgewater – 29, 39.
Bristol and Exeter Railway – 14, 21, 23, 24, 26, 48, 50, 51, 73.
Bristol (incl Temple Meads) – 19, 27, 50, 182, 208, 210.
British Railways – 44-48,
British Railways Modernisation Plan – 211.
Brotherhood and Hardingham – 121.
Brunel, Isambard K – 24, 44.
Bulleid, Oliver V S – 49, 215.
Burnham – 6, 15, 24, 81.
Bunham accident (1914) – 148.
Bury, Curtis & Kennedy – 63.
Bury Edward & Co – 63.

Chambers, Herbert – 154, 162, 175, 180.
Christian, C – 73, 83.
Churchward, George – 41, 96, 131, 134, 136, 137, 143, 156, 178.
Clarke, D H – 60.
Clayton, James – 41, 124, 125, 131, 132, 137, 144.
Clayton, Thomas – 77,
Cleaver, C F – 208.
Cole – 15, 52,
Coleman, Tom – 49, 154, 174, 175, 179-181, 186, 203, 207, 208, 221.
Collett, Charles – 158, 178, 208, 225.
Colson, Alfred – 30, 32, 81.
Combe Down Tunnel – 19, 73, 167-170, 186.

Corfe Mullen – 37.
Court of Chancery – 20, 67, 68.
Cox, A E – 208, 210.
Cox, Ernest – 158, 160, 178, 203, 204, 218.
Cudworth, James – 64-66.

Deeley, Richard Mountfield – 8, 49, 71, 75, 118-135, 144, 145.
Derby – 31, 32, 76, 81, 96, 98, 100, 105, 106, 108, 110, 111, 114, 115, 119, 124, 128, 129, 132, 134, 137, 140, 141, 145, 148, 151, 159.
Devonshire Tunnel – 19.
Dorset Central Railway – 15-17, 48, 51, 55-57, 73.
Drummond, Dugald – 191, 194.
Dubs & Co – 191, 195.
Duck, Son & Pinker – 124.
Dykes, Robert Armstrong – 26-27, 30-32, 37, 40, 81, 148, 149.

East &West Junction Railway – 58.
Eastern Counties Railway – 53.
Eclipse Peat Co – 209.
England, George – 48, 57-64, 67, 70, 85, 86, 178.
Evercreech – 30, 135, 214, 223.

Eyre, George – 149.

Fairburn, Charles – 221.
Fenton, Craven & Co – 77.
Firbank, Joseph – 63.
Fisher, Benjamin – 22, 30, 69, 80, 85, 93, 108, 109.
Flamme, Jean Baptiste – 138.
Fowler, Henry – 41, 49, 75, 96, 119, 125, 128-135, 138, 144, 150, 155, 157, 159, 160, 162, 170, 175, 180, 185, 186, 188.
Fowler, John – 70, 73, 83, 97, 98.
Fox, Walker Co – 48, 58, 69-73, 97, 103, 104.
French, William Henry – 30, 108, 109, 114.

Garbe, Robert Dr – 138.
Gass, Edward – 156.
Gauge of Railways Act (1848) – 57.
Glastonbury – 14, 15, 22, 50, 208, 223, 224.
Gooch, Daniel – 24, 53.
Gooch, John – 49, 53.
Granet, Guy – 128, 129.
Great Central Railway – 122, 137, 140.

**Fox Walker** 0-6-0 saddle tank No.5 built in 1875 photographed at Radstock in the same year, or so it seems if the date on the back of the old, faded print is accurate. (D. Neal)

Great Eastern Railway – 82,
Great Exhibition (1851) – 59.
Great Northern Railway – 137, 138
Great Northern & Midland Railway – 27.
Great Western Railway – 15, 22, 24, 30, 50, 53, 57, 73, 88, 131, 136, 137, 140, 151, 156, 178, 208, 223-228.
Gregory, Charles – 22, 57.
Gresley, Herbert Nigel – 130, 138, 143, 156, 165, 196.
Guy, Peter – 223.

Hardy Railcar Co – 208.
Hatcham Iron works – 59, 60, 64.
Hawksworth, Frederick – 226.

Hawthorn & Co – 88, 226.
Henstridge – 14, 27, 28, 30, 39, 40, 43, 45, 48, 52, 63.
Henstridge accident (1944) – 198, 199.
Highbridge (incl workshops) – 14, 27, 28, 30, 39, 40, 43, 45, 48, 52, 63, 105, 107-112, 116, 119, 124, 135, 137, 140, 141, 143, 147, 149, 158, 166, 173, 178, 222.
Holden, Stephen – 196.
Hughes, George – 151, 155, 156, 160, 162, 177, 186.
Hunslet Engineering Co – 88.
Hydraulic Engineering Co – 121.

Illustrated London News – 49.
IMechE – 145.

International Exhibition (1862) – 60, 62.
International Rail Congress (1954) – 218.
Ivatt, Henry – 200, 207.

Jarvis, Ron – 49, 221.
Jennings, J H – 168, 169.
Johnson, Samuel -36, 75-78, 81-85, 88, 89, 92, 93, 96, 98, 101, 103, 105, 107, 111, 114, 118, 121, 127, 131, 138, 140, 145, 151.
Jowett, Daniel – 74.

Kerr, Stuart – 98, 181.
Kirtley, Matthew – 31, 75-78, 81, 98, 99, 105, 119.
Kitsons – 57, 81.

**2P 4-4-0** No. 44, built in 1928, stands at Evercreech Junction a year later with the down Pines Express. This engine remained in service until November 1959. (R.Hillier)

**Locomotive types:**
ARLE 2-8-0 (proposal) – 150, 151.
Aspinall Class 2/3 4-4-0 – 130.
    Class 5 – 2-4-2T – 130.
    Class 27 0-6-0 – 130.
    Class 30 0-8-0 – 130.
Bagnalls 0-6-0T – 233.
British Railways Class 3MT
    2-6-2T – 231.
    Class 4 2-6-4T – 227-231, 234, 236.
    Class 4 2-6-0 – 219, 221.
    Class 4 4-6-0 – 220, 221.
    Class 5 4-6-0 – 217, 218, 227-229.
    Class 8P 4-6-2 – 205, 233.
    Class 9F 2-10-0 – 205, 213,
        227-229.
    Class 35 Hymek Diesel – 233.
    DMU – 210, 211.
    Sulzer Type 4 Diesel – 233.
Bury Edward & Co 2-2-2T – 63.
Bison Class – 0-6-0 – 53, 56.
    0298 Class – 2-4-0 – 53.
    1528 Class 0-4-0T – 122.
Deeley Class 1 – 0-4-4 – 76.
    Class T7 - 0-4-2 – 78.
    134 Class – 0-4-4T – 78.
    200 Class – 4-4-0T – 78.
    417 Class – 0-6-0 – 78, 79.
    477 Class – 0-6-0 – 78, 79.
    990 Class 4-4-0 – 173.
    2000 Class – 0-6-4T – 122,
        132, 133.
    4-6-0 (proposal) – 127, 128.
    0-8-0 (proposal) – 127, 128.
    4-4-0 – 126, 127.
Drummond Class 700 0-6-0 –
    194, 195.
Fox Walker &Co – 0-6-0 – 70-73,
    104, 127.
Garratt 2-6-0 + 0-6-2 – 156, 175, 186.
George England – 2-4-0/2-4-0T –
    58-64, 85-87,
Great Eastern Railway S69/B12
    4-6-0 – 196, 197.
Great Western Railway Class 56xx
    0-6-2T – 227.
    Class 302 0-6-0 – 224.
    Class 850 0-6-0ST – 224.

Class 2215 0-6-0 – 223.
Class 2251 0-6-0 – 223, 224.
Class 2721 0-6-0 – 224.
Class 2800 2-8-0 – 41, 131, 136.
Class 5700 0-6-0 – 223-225.
Class 9400 0-6-0 – 223, 226.
Castle Class 4-6-0 – 233.
Great Bear 4-6-2 – 131.
Manor Class 4-6-0 – 233.
Saint Class 4-6-0 – 131
    Diesel Railcar – 208-211.
Hawthorn Class 2-4-0 – 88.
Hercules Class 2-4-0 – 53, 54.
Hughes 4-6-2 (proposal) – 156, 162.
    2-8-2 (proposal) – 156, 186.
John Fowler 0-6-0 – 19, 69, 70, 73, 83.
Johnson 4-4-0 – 79, 83, 114-116,
    118, 121.
Johnson Class 483 4-4-0 – 137-141,
    151, 156, 175.
Johnson 0-4-4T – 82, 88-93, 95-97.
Johnson 0-6-0 – 19, 102.
Johnson 0-4-4ST – 8.
Johnson 3-cylinder compound
    4-4-0 – 118, 121, 158.
L&M 2-8-0 – 136.
L&Y Class 21 0-4-0T – 173.
LMS – 2P 4-4-0 – 156, 157, 175, 176,
    177, 181, 184, 185, 196, 204.
    2MT/4MT 2-6-0 – 205-207.
    2MT 2-6-2 – 207.
    3P 2-6-2T – 174, 185-187.
    3F 0-6-0T – 156, 170, 171, 178, 208.
    4P 4-4-0 – 157, 158.
    Black Five 4-6-0 – 10, 154, 174,
        177, 179, 181, 182-185, 193, 203,
        217, 218, 228, 233.
    7F 0-8-0 – 156, 160.
    8F 2-8-0 – 10, 42, 45, 154, 159, 174,
        186, 188, 189-191, 227, 232, 233,
        235-237.
    Jubilee 4-6-0 – 180, 184-186.
    Lickey Banker 0-10-0 -187.
    Patriot 4-6-0 – 185.
    Princess Royal 4-6-2 – 180.
    Royal Scot 4-6-0 – 156, 175
    10000 Class Diesel – 208.
LNER – B1 4-6-0 – 218.

    O1 2-8-0 – 137, 138, 143.
    P1 2-8-2 – 162.
LNWR G1/G2 0-8-0 – 160.
    Prince of Wales Class 4-6-0 –
        157, 158.
LSWR K10 4-4-0 – 194.
    S11/V11 – 193, 194.
    T1 4-4-0 – 192, 194.
    T9 4-4-0 – 191.
Locomotion No 1 – 165.
Mazeppa Class 2-2-2 -53.
Midland Railway 4F 0-6-0 – 144,
    150, 153, 166, 198, 213.
    7F 2-8-0 – 9, 11, 41, 44, 102,
        141-149, 154, 159, 162-166, 169,
        181, 186-188, 195, 199, 201, 202,
        223, 238.
Neilson, Reid & Co 2-4-0 – 78, 104,
    208, 209, 233.
Nelson Class 2-4-0 – 56, 57.
Paget 2-6-2 – 121, 123, 125.
PRR 2-8-0 (H6) – 135, 136.
Prussian Rly Class S4 4-4-0 –
    138, 139.
Riddles Austerity 2-8-0 –
    223, 227.
Robinson 2-8-0 – 137, 140.
Rothwell 4-2-4T – 51, 52.
    4-4-0 – 50, 51.
Sentinel 0-4-0T – 172, 173, 233.
Slaughter, Gruning & Co
    0-4-ST – 111.
Southern Railway 4-6-2 – 45,
    212-215, 232, 233.
    H15 4-6-0 – 140.
    Schools Class 4-4-0 – 214, 216.
    U/U1 Class 2-6-0 – 214.
    Z Class 0-8-0 – 216.
Sussex Class 2-2-2 – 55.
Vulcan 2-4-0 – 66, 67, 130
Vulcan 0-6-0 – 104.

Lambert, Henry – 162.
Lancashire and Yorkshire
    Railway – 130, 151.
Lancaster and Heysham
    Line – 122.
Lawrence, Charles – 156.

**Now with** LMS insignia Fox Walker 0-6-0ST, No. 1507, nears the end of its life after nearly 55 years of service. This engine was purchased in 1876 and withdrawn in December 1930. Here the saddle tank is captured undertaking shunting duties with an unidentified 7F in the background. (R Hillier)

Lehigh & Mahanoy Rail Road – 136.
Lemon, Ernest – 155, 159, 179.
Locomotive Magazine – 59, 181, 182, 188.
Loder, Sydney – 169.
London, Midland and Scottish Railway – 10, 40-43, 48, 96, 104, 130, 153-155, 159, 162, 166, 173, 178.
London Midland Region – 203.
London & North Eastern Railway – 156, 162, 172, 196, 197.
London & North Western Railway – 122, 123.
London and South Western Railway – 15, 18, 22, 23, 29, 32, 40, 46, 48, 51-53, 57, 59, 73, 86, 82, 106, 117, 133, 134, 140.

Masbury – 20, 73, 182.
Mason Science College – 129.
Maunsell, Richard – 132.
Mendip Hills – 20,
Midland Counties Railway – 74.

Midland Railway – 18, 22, 23, 30, 32, 35, 36, 40, 45, 66, 68, 73-81, 84, 90, 93, 97, 98, 101, 103, 110, 111, 115-117, 121-125, 128, 130, 145, 151, 153, 159, 160, 162.
Mitchell, James – 98.
Morecombe & Heysham Line – 122.
Motor Car Co – 125.

Neilson & Co – 82, 98-100, 103, 153.
Neilson, Reid & Co – 78, 107, 195.
Neilson, William – 98.
Nine Elms – 52, 54, 55, 57, 86, 191, 192, 194.
Norman, Isaac – 169.
North British Locomotive Co – 175.
North British Railway – 78.
North Midland Railway – 74.

Paddington – 15, 24.
Paget, Cecil – 119-123, 125, 128, 130
Paget, George Ernest – 119, 121, 128.
Paris-Orlean Railway – 166.
Pearce (Fireman) – 168, 169.
Pearson, James – 49-53, 57.
Pecking Mill accident – 66.
Pennsylvania Rail Road – 133-136.
Peppercorn, Arthur – 218.
Peters, Ivo – 9, 214, 217, 218, 226.
Pike, Peter – 10, 216-228, 226.
Pines Express – 43, 185, 217, 228, 229.
Poole and Bournemouth Railway – 22.
Prussian Railway – 138.

Radstock – 24, 38, 45, 73, 83, 95, 111, 113, 145, 154, 222.
Radstock accident (1876) – 25-27, 80.
Rail accident (Feb 1877) – 89, 92.
Railways Act 1921 - 40, 160.
Railway Centenary Exhibition (1925) – 165.
Railway Executive Committee 40, 148.
Railway Hotel Bridgewater – 14.

Railway Magazine – 28
Ramsbottom Safety Valve – 162, 177.
Read, Robert Arthur – 21, 22-25, 62, 81.
Redhead, Thomas – 151, 154, 156.
Redman, A S – 156.
Reid, James – 100.
Reid, Robert Whyte – 36.
Riddles, Robert – 49, 180, 203, 204.
Robinson, John – 122, 143, 151.
Ross & Co (Ross pop safety valve) – 162, 167, 177.
Rothwell & Co – 50, 51.
Ryan, Mervyn – 40, 133-138, 142, 145.

Sacre, Charles – 78.
Schmidt, Wilhelm Dr – 138, 140, 144, 151, 162.
Scott, Archibald – 22, 24.
Scovell, Ted – 223.
Sealey, S – 43.
Sentinel – 170, 172.
Sharp Stewart – 195.
Shepton Mallet accident (1885) – 94.
Slaughter, Gruning & Co – 111.
Slessor, Frederick – 22, 69, 80, 85.
Smith, Peter – 9, 192, 193, 195, 227.
Somerset Central Railway – 13-17, 27, 48, 50-52, 57-59, 73.
South Drain accident (1949) – 208, 209.
South Eastern & Chatham Railway – 131, 194.
South Eastern Railway – 63, 64.
Southern Railway – 11, 40, 43, 132, 151, 153, 178, 191-194, 197.
Southern Region – 203, 212, 214.
Spiller, Ron – 223.
Stanier, William – 49, 155, 159, 178-181, 186, 188, 203, 221, 232.
Stenson, William – 74.
Stephenson, George – 74, 76.
Stephenson, Robert – 57, 74, 163.
Stirling, Patrick – 89.
Stockton and Darlington Railway – 74, 77.
Sturminster Newton – 39.
Sturrock, Archibald – 78.
Swindon – 30, 136, 218, 221, 224, 229.

Symes, Sandham John – 124, 125, 131, 137, 160.

Taff Vale Railway – 69.
Templecombe – 10, 16, 19, 29, 31, 37, 52, 83, 89, 185, 206, 207, 214, 220, 221, 235.
Thom, Robert – 196.
Thompson, Edward – 218.
Titfield Thunderbolt – 8, 45.
Transport Act (1947) – 199.
Trench, A Colonel – 168, 169.
Turner, Thomas Professor – 130.
Tyer, Edward – 37.
Tyler, H W Captain – 92.
Tynings Bridge – 113, 172.

University College Nottingham – 133, 145.
Urie, David – 179.
Urie, Robert – 133, 140, 191.

Vulcan Foundry – 6, 43, 48, 66, 67, 93, 98, 100, 102, 103-105, 129.

Walkers, T C – 20, 69.
Waterloo – 15.
Webb, Francis W – 89.
Wellow – 43,
Wells – 15, 30.
West Flanders Railway – 64.
West Pennard accident (1958) – 223.
Western Region – 211, 222, 227.
Wheeler, G H – 156.
Whitaker, Alfred - 30-36, 39, 40, 109-111, 117, 118, 122, 124, 126, 128, 132, 134, 135, 144, 145.
Whitaker, Alfred Henry – 149.
Willen Valve System – 121.
Wilson, E B & Co – 50, 77.
Wimborne – 44.
Windle, Edward – 218.
World War 1 – 40, 41, 47, 109, 122, 131, 134, 145-149, 162, 196.
World War 2 – 44, 139, 190, 195-199.
Writhlington – 38, 42, 233.

**Towards the** end of the S&D's life a local stopping train pulled by BR Standard Class 4MT 4-6-0 No. 75072 pauses at Radstock Station, which displays few signs of life. (Author)